Diggin g up Bones

Digging up Bones

The excavation, treatment and study of human skeletal remains

D R Brothwell
Institute of Archaeology,
University of London

Third edition

British Museum (Natural History)

Oxford University Press

British Library Cataloguing in Publication Data

Brothwell, D.R.
  Digging up bones.—3rd ed
  1. Excavations (Archaeology)   2. Anthropology
  I. Title
  930.1′028′5   GN70

ISBN 0-19-858510-1 (Paperback)
      0-19-858504-7 (Case)

Third edition first published 1981 by

British Museum (Natural History)
Cromwell Road, London SW7 5BD and

Oxford University Press, Walton Street, Oxford OX2 6DP

OXFORD LONDON GLASGOW NEW YORK TORONTO MELBOURNE AUCKLAND
KUALA LUMPUR SINGAPORE JAKARTA HONG KONG TOKYO DELHI BOMBAY
CALCUTTA MADRAS KARACHI IBADAN NAIROBI DAR ES SALAAM CAPE TOWN

Printed in Great Britain by BAS Printers Ltd.

Contents

Contents continued

Introduction

Bones are still commonly a problem to archaeologists, even though the human skeleton offers a no less fruitful subject of inquiry than ceramics, metals, architecture or any other field of historical or prehistorical study. It can provide much information on human societies of the past, and in fact no social reconstruction can be complete without examining the physique and health of the community. Osteological material may also provide interesting information of a more specialized nature, such as the effects of the environment on modern and earlier populations, or the evolution of diseases.

Unlike some archaeological procedures, excavation and removal of bone material, if carried out with due care and discipline, need in no way be a destructive process. Indeed, in some cases, the excavation of skeletal material ensures its having a longer life than it would otherwise be guaranteed in the soil.

Modern physical anthropology has tended to concern itself more and more with variability in populations, by means of the study of the physical characters of the living. The detailed analysis of human blood, for instance, is a promising field of research, for a number of different biochemical characters associated with it are inherited in a fairly simple way, and do not change in composition or reaction throughout life. Few of the other characters which are as yet known are so straightforward in application. The analysis of pigmentation (the colour of skin, eyes and hair), although presenting difficult problems as regards its inheritance, nevertheless has far fewer variables to contend with than the skeleton. Of those characters which can be studied in bone, nothing is as yet known of the number of genes determining bone shape and size, and there is still much to be learnt about the influence of diet, climate, injury and disease. Some of the individual features of the blood can be studied by using those parts of the bone which have contained red marrow. In spite of the limitations in the study of skeletons, there is little else left of earlier man, for the mummies and desiccated bodies that have survived are too few to be of much value. We must therefore try to obtain as much reliable information as possible from the bones.

Much of the research on early remains of man is concerned with the following questions: Where did early man originate and what was he like? How variable has man been from prehistoric to recent times, and how ancient are the distinctions between present varieties of man? To what extent is the variability in earlier groups determined by the *genotype* or inherited gene complex of the individual, and how much is governed by environment? Are any features restricted to particular geographical or climatic areas, and might they be associated with some 'adaptive' advantage in the groups possessing them? Which diseases have affected him throughout his evolution, and which have attacked him in more recent times? Finally, how many years could each individual expect to live during those earlier times?

This handbook cannot give the detailed answers to these questions, but rather it tries to describe techniques and lines of inquiry that may in the long run present us with some of the solutions. It should not be thought that because only one skeleton is discovered, that it is not worth reporting, for provided it is well dated, it helps to increase the number of specimens available for study. In Britain alone during the past century many skeletons have been thrown away, sometimes without even a preliminary note in a local archaeological journal. The result is that we are still sadly ignorant of our predecessors, especially as regards their regional variability and their affinities with other regional samples.

'Specialists' are not always available to report on excavated bones. This has sometimes resulted in bones being left for two or three years or more before any study of them has been made. The archaeologist may even have to choose between briefly reporting on bones himself, or waiting a considerable time before a 'specialist' can undertake an examination. In any case, an elementary knowledge of the skeletal remains that are constantly being revealed in excavations is now essential to all archaeologists if they are to evaluate and deal efficiently with material found on their sites.

A detailed description of human remains will be usually beyond the scope of the general archaeologist, but a provisional one, suitable for publication with the other excavation details, should not be. What sort of information should be recorded? This is a problem which also faces the 'specialist'. Miss M. L. Tildesley (1931) began to answer this question, but no final assessment of priorities and the range of data to be recorded has ever been attempted, although there are various publications on bone remains from a forensic point of view. Reference to previous reports does not help greatly, as the arrangement and degree of detail varies considerably. Some suggestions on the standardization of osteological reports therefore seem long overdue and one or two pages are devoted to this topic.

It is hoped that what follows may be of interest and value to students of general anthropology and anatomy, as well as to the archaeologist and the layman.

It may be as well to emphasize here that the attainment of a basic knowledge of the human skeleton is not an exhaustive process, and is quite within the reach of those without anatomical training, whether adult archaeologist or schoolboy amateur. First-hand acquaintance with bones is advisable, but this should not present any insuperable difficulty, for most museums have some such material.

Differences between human and animal bones have already been dealt with by Cornwall (1956) and Schmidt (1972) and will receive only brief consideration here (p.36).

Those wishing to have a better comparative knowledge of animal bones are also referred to Ryder (1969), Sissons & Grossman (1953) and Boessneck (1969).

I Notes for guidance in excavating and reporting on human remains

1 The site of discovery

If the notes on the human remains are to form part of a larger excavation report, then a description of the site will be dealt with sufficiently in the other sections; but if the bones are to be discussed in a separate publication, some details of the district and site of discovery should be given.

In reporting important discoveries, a map should always be included, showing the nearest town and general topographical features. If this is not possible, the National Grid-Reference, nearest town and name of the immediate vicinity to the excavation should be given. If, for some reason, the material is not receiving publication but a report is required for filing, it is nevertheless still worth noting the area of the finds in detail; and if record sheets are being used to catalogue the material, it is a good idea to put a sketch map on the back, as Stevenson (1930) suggests.

2 Place of burial

The environment in which a skeleton is found may help to solve certain problems. Type of barrow may denote the period when buried; a large number of jumbled skeletons in a pit might suggest plague victims, while the isolated burial of a skull, as in the Palaeolithic calvarium from Whaley Rock Shelter, Derbyshire, could perhaps indicate a ritual burial. It is therefore important to note the place of burial as accurately as possible, and though of course full reports are undertaken in official excavations, it sometimes happens that human bones are given to museums without sufficient description as to their origin. The following are some of the main situations in which bone material is to be found.

a) Various geological deposits. River gravels, sand, peat bogs, shell mounds, volcanic ash, calcareous deposits.

b) Under burial earthworks. As primary, secondary or tertiary burials.

c) Associated with early building structures. Rampart ditches, under floors, in wells.

d) Burial on or near the battlefield. Slain during or after battle. (The mutilations may show that the wounds were received while fighting, while injuries to the neck region and positioning of the arms may point to later execution.) The medieval Japanese burials from Zaimokuza (Suzuki et al., 1956) are a particularly good example of roughly buried battle victims, and of over 500 individuals, many displayed sword injuries.

e) Plague pit, or jumbled secondary burials associated with the disturbance of a plague cemetery.

f) Ordinary cemetery burials. With or without definite rows and orientation.

g) Incidental burials. For example, fragments in a hearth or refuse pit.

h) Cave or rock-shelter. Either the intentional burial of the whole or part of a body, or as bones selected for ritual purposes. In the case of the Neanderthal Monte Circeo cranium, it was found in its original 'ritual' position, just as before the landslide had sealed off the cave entrance (Brodrick, 1948).

3   Types of burial

As Goodwin (1945) points out, burial rites are hide-bound by custom, and the position and orientation of the body may help to show the distribution of a cultural group, both in space and time. Variations in type of burial associated with a single people also help to establish divergence of belief and custom, although, as Ucko (1969) has pointed out, ethnographic interpretation from burials can be a hazardous business. That there were considerable variations in burial type even during prehistoric times is well exemplified by Marija (1956), but interpretation is not always easy. Harrisson (1957) discovered five fairly distinct types of burial in the Great Cave of Niah, Borneo. Mukherjee *et al.* (1955) were able to show that there had been some changes in the burial custom of the people of Jebel Moya in relation to the period of settlement. By applying the x^2 test to the frequencies of burial types, he was able to show: that the form of burial changed in time, the extended type being more common in later times; the hand(s)-to-face burial became more common with the growth of the settlement; the supine position is more often associated with an extended burial, and the position on the side with a flexed type of burial.

Although orientation of the body, to face east or west, may sometimes be valuable information, especially in deciding whether a cemetery is pre-Christian or Christian, it may nevertheless lead to misinterpretation.

The main types of burial encountered in the field may be summarized as follows:

a) Fragmented remains, either loose or contained in an urn.

b) Extended burial. The orientations in a group of burials may or may not be similar. Special positioning of arms or head sometimes occurs.

c) Flexed burial. Usually the body has clearly been laid on one side, with the arms and legs bent but generally in no special positioning.

d) Contorted burials. There are numerous atypical positions into which the body may be placed. Such abnormal postures can denote hurried burial, or the burial of battle victims (perhaps already stiffened by rigor mortis).

e) Other burial types. It seems possible that other positions, not of a grotesque variety, may be encountered. Certain African groups, for example, tightly bind the corpse so that it takes up a position fitting to the dead man's job in life; and such positions may remain to some extent even after decomposition.

4   The process of excavation

When an area of bone is revealed in the earth, the exposed margins should be gently cleared away with a small pointing trowel or knife and a brush. If it is a long-bone (e.g. femur, tibia), it is important to see if any bone adjoins it or articulates with it. In the case of ribs, try to find more in line. Exposure of a few bones should enable the side (whether left or right) and area of the body to be tentatively ascertained, as well as giving a rough idea of the posture of the body. Knowing these facts, the excavator will be able to clear away the rest of the soil with a little more confidence as to what bone will be next. It should be noted that the articulation of one or two bones need not necessarily denote a complete or articulated skeleton. Battle mutilation or post-mortem decomposition before or after burial may give rise to incomplete or deranged skeletons. After exposing the bones, it is a good policy to let them dry out slowly in normal (but not intense) sunlight, which should harden them appreciably. The final cleaning before removal may be left until this has been done, particularly if the bone has shown itself to be easily chipped and scratched. A fairly stiff brush should be used to clean away earth directly on bone, a small pointing trowel or knife being used to loosen it.

Bones should be liberated as much as possible from the surrounding earth before removal. It is very tempting to remove, say, a long-bone, when only a part of it is uncovered, but this may cause breakage of a brittle bone, especially if the earth is stony. It is not usually necessary to attempt to raise skeletons, or parts of skeletons, within blocks of containing earth. However, in cases where the bones are particularly brittle and ancient, then it may be preferable. Woolley (1949) for instance, found it necessary to transport the brittle skeletons from Ur in protective blocks of wax.

Also, if the skeletons are in a solid calcarous matrix, as in certain Upper Pleistocene (Mount Carmel) and australopithecine remains, they need transporting to the laboratory *en bloc* where slow drilling and chipping, or dissolving of the matrix in acetic acid, will be demanded.

When removing the skull from ordinary grave earth, it is advisable to undermine the underneath surface of the cranial vault and face. Do NOT attempt to lift the skull when it is only half exposed, as this inevitably results in the ramus, temporal and zygomatic regions, if still in the soil, breaking away. If the mandible can be removed before the cranium, then do so, for this gives the excavator more freedom (often it drops away from the cranium on decomposition). The excavator must try to lift and support the face and vault at the same time; if not, earth in the nose and sinuses may cause it to break away or to fracture. The positions of the hands will depend upon how the skull rests, and no routine position can be described. However, in the case of the face, support at the teeth is preferable. Do NOT lift with the fingers over the fragile margins of the nasal (pyriform) aperture.

The orbits may be cleaned before removal if the posture allows, but this demands great care as parts of the walls are only of paper-like thickness (especially at the lacrimals). Loose teeth should be boxed and kept with the skull.

Leechman (1931) suggests that before removing the skulls, the ear passages should be plugged with cotton wool or other suitable material, so that the auditory ossicles will not be lost. However, it is usual in British material to find earth in the auditory meatus, so that no extra precaution need be taken. Comparative osteometric work can be achieved even on these minute bones (Masali, 1964).

The hyoid cartilage should be looked for and if found retained with the skull. The laryngeal (throat) and costal (chest) cartilages are sometimes calcified, in which case these should also be saved. Other kinds of calcified structures may at times occur. Weiss and Møller-Christensen (1971), for instance, report on the presence of multiple calcified *Echinococcus* cysts in the abdominal region of a medieval Danish female.

5   Samples for analysis

As archaeological specializations become more developed, so the need to obtain samples for micro-analysis becomes more necessary. A number of the techniques now in use are applicable either directly or indirectly to skeletal material.

Bone samples for blood grouping

It may eventually be possible to compare certain early populations by the ABO group reaction of bone. It is doubtful whether the Rh group, as Gilbey & Lubran (1953) suggest, or indeed any of the other blood groups, will ever be widely ascertainable, owing to the greater instability of these organic substances. Experiments in determining blood group substances in ancient bones were begun by Boyd & Boyd (1933) and extended by Candela (1936) and Matson (1936). More recent work includes that of Laughlin (1948), Thieme and colleagues (1956, 1957a, 1957b), Gray (1958), Smith (1959, 1960), Kout, Vacikova & Stloukal (1965), Borgognini (1968), Borgognini & Paoli (1972), Lengyel (1971). A number of problems still remain in these studies of palaeoserology, particularly as regards reactive substances in the soil and bacterial action on the antigens (the organic substances in blood, bone and other tissues that produce the blood-group reactions).

In the case of well-preserved and fairly complete human remains, the researcher will be able to obtain his own bone samples in the laboratory. However, where preservation is poor, the excavator should leave at least smaller fragments of post-cranial material free from preservatives, pending future chemical or serological analysis. Ideally, a bone or area of bone containing clean cancellous tissue (as in the top of the femur) should be placed in a polythene bag to await analysis. If the excavator finds bones very badly crushed, or in a poor state of preservation, he should note that even though it is not worth making a special effort to remove and reconstruct the fragments, they may still be useful as bone samples.

Organic matrix of bone: nitrogen analysis

Another group of organic substances, the amino-acids, have been found to be stable in bone over relatively long periods (Abelson, 1954). Chromatography of hydrolized bone shows these acids to be present in varying strengths, differing in bones from different sites. These amino-acids are part of the organic matrix of bone (i.e. bone protein or collagen). Even quite small fragments of human bone, if older than Neolithic, may be worth preserving to test for organic content. For example, a small splinter that was detached from the Neanderthal mandible 'Monte Circeo II' was used to prove the presence of histologically demonstrable organic matrix (Ascenzi, 1955). Estimation of the nitrogen content of bones is one means of assessing their amino-acid or protein content. By this method of analysis Heizer & Cook (1952) showed that bone protein ('ossein') in bones preserved in similar conditions, disintegrates at a fairly constant rate and is therefore useful in relative dating. This method has since been applied to a number of remains including the Piltdown fragments (Weiner, Oakley & Clark, 1953).

Radiocarbon dating of bones

When adequate quantities are available, the residual protein in bones can sometimes be dated by the radiocarbon (C^{14}) method. This expensive technique has only occasionally been applied to human bones, largely because in the case of important specimens it would be difficult to spare sufficient material for the destruction that the method entails. It was applied to a few grams of the Piltdown skull, which was thus shown to be probably more than 600 years old (de Vries & Oakley, 1959). The method has also been applied to a larger sample (100 grams) of the Galley Hill skeleton, which was thereby dated as 3310 ± 150 years bp (Barker & Mackey, 1961). More recently C^{14} dating of pathological skeletons has assisted the establishment of a pre-Columbian date for treponemal disease in the New World (Brothwell & Burleigh, 1975). Radiocarbon dating is more precise when applied to wood or charcoal: if these are associated with a human skeleton they may be used as an indirect means of dating the bones. Completely calcined bones cannot be dated by the radiocarbon method, although if only charred they sometimes can be.

Samples for inorganic analysis

Fragments are valuable in the relative dating by fluorine analysis or radiometric assay, which may help to show chronological ambiguity as in the Galley Hill skeleton Oakley & Montagu, 1949) and the Piltdown remains (Oakley, 1955a; 1955b) or contemporaneity as in the *Homo erectus* skull and femur from Trinil (Bergman & Karsten, 1952).

There is now a growing interest in the analysis of bone for other trace elements, not for their relative dating value, but in relation to aspects of the life of individuals. See, for instance, the general survey by Lambert, Szpunar and Büikstra (1979). Lead in particular has so far received attention, and there is some evidence that the lead content of bones may be evidence of lead pollution in earlier societies (Grandjean & Holma, 1973; Waldron, Mackie & Townshend, 1976, Waldron 1973).

The possible association of certain trace elements with diet has also resulted in a number of other exploratory investigations (Gilbert, 1977; Szpunar, Lambert & Büikstra, 1978; Lambert, Szpunar & Büikstra, 1979; Schoeninger, 1979).

This is clearly a very promising research field, in which considerable work will take place.

Persistence of hair on skulls

Although not directly concerned here, it may be noted that in some climates, particularly those leading to the desiccation of the remains, hair may still be found on the skull, and it is therefore advisable to bear this in mind when clearing the earth away from the vault. Hirsch & Schlabow (1958) noted hair fragments of a Bronze Age date, and even the climate in Britain allows hair to remain in certain conditions. Mortimer (1905), for example, noted the discovery of hair in four pre-Saxon burials and body hair was found on a number of Roman skeletons at Poundbury, in southern England. Any such hair, however small in quantity, should be carefully removed and placed in an airtight, or at least moth-proof container, as it can provide a variety of information about pigmentation and structure (Brothwell & Spearman, 1963).

Soil samples and bones

It is worth while taking small samples of the earth directly on, or even within the bones of a skeleton. These will enable soil conditions to be easily ascertained by a future worker, and may be of value in checking blood-group reactions on the bones. It may also provide material for pollen analysis.

6 Photographing bones in the field

Cookson (1954) and others have discussed the photography of bones in excavations, but certain points are worth repeating here.

An important problem is how to clean the bones *in situ* sufficiently to give a clear picture. Great care should be taken not to disturb the original position. It is worth recording as many burials by this method as possible, but one exposure is usually sufficient.

After clearing as much soil off the bones as possible, they may be washed once or twice with a panel brush and clean water, as this will make them much lighter in colour. Care should be taken in clearing soil from within the orbits, ribs and pelvis. If any bones are loosened, the use of lumps of concealed plasticine will enable them to be retained in position.

The selection of light-filters may be advisable in the case of certain types of bone staining and earth colour, although this is not always necessary. The attitude of the skeleton should not be obscured through bad positioning of the camera, although photography may be to some extent restricted by the form of the trenching. A scaling rod should be visible at the side of each view photographed. Ritchie & Pugh (1963) have shown that, when dealing with badly decayed skeletal remains, ultra-violet fluorescence photography is valuable in revealing bone detail. Indeed, by this technique it is sometimes possible to confirm the original occurrence of an inhumation (Plate 1.1) by the production of a soil 'silhouette' of the interred, but now decayed, corpse (Brongers, 1966).

7 Labelling and numbering

Nothing is more frustrating to future workers than having to deal with remains that are badly numbered or not numbered at all. Although the form of cataloguing may differ with the type of excavation and museum or institution in which the bones will eventually be kept, where possible the date of discovery should be included as well as the grave number.

The skull and at least some post-cranial material should be numbered with Indian ink on removal from the site. Ordinary ink should not be used as it will run on washing. An indelible pencil is particularly useful for marking bones while in the field. Most deposits and kinds of preservation allow bones to be numbered even without washing. For removal to the laboratory, deposits within skulls should be taken out as far as possible. Even when the numbers can only be provisional, they should nevertheless be allotted, for the risk of mixing material is always considerable. The symbols should be placed in fairly inconspicuous places, and need not be more than 5 mm high. The name of the site and period can also be put on the skull. If it is found that the bone surface is particularly rough after cleaning, then a small area can be smoothed with a suitable cement or adherent substance which will allow lettering to be superimposed.

Labels can also be placed with the bones, or numbers put on the containers. Gummed labels should not be stuck on the skulls or other bones.

Plate 1.1 An ultraviolet fluorescence photograph of a soil silhouette of a buried corpse from Holland (courtesy J. A. Bongers).

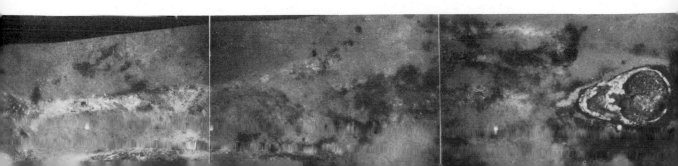

8   Packing and transport

When packing bones for removal, strong tissue paper or even newspaper is more useful than cotton wool. Cardboard boxes are more easily obtainable and better to handle as containers. Old grocery and stationery boxes are ideal for transporting bones to the museum or laboratory if a private vehicle is to be used, but strong wooden boxes will be needed if they are to be sent by train or ship.

In the case of fairly complete skeletons, the bones of the skull, vertebral column, ribs, hands and feet, should be packed separately when possible, as this helps considerably with the sorting and subsequent description. In the case of a child's skeleton, with the less defined shapes and incompletely formed bones, a special effort should be made to keep separate the various areas.

Most bones need no special positioning in the box, and it is only necessary to ensure that fragments do not rub against one another. In the case of the skull, however, the position in transit depends upon its condition. When possible, the mandible should be wrapped separately. If the cranium is in good condition, then it can be placed on paper with the teeth and base downwards. However, when the face and/or base is broken and liable to fall away, the vault can be placed on crumpled paper and supported with more at the sides.

9   Soils and preservation of bones

The kind of preservation of bone not only varies considerably from soil to soil, but also from one area of burial to another, through minor soil differences. Knowledge of the soil may help the preparator to decide upon the most appropriate method of cleaning and preserving the bones. Even before seeing the bones it is sometimes possible to infer their probable preservation. The chalky soil of Dover, for example, usually leads to early loss of organic matrix so that, for example, Saxon bones are brittle and porous, whereas in the clayey earth of the Langwith Cave, Derbyshire, human bones of much greater antiquity show a remarkable degree of preservation, appearing almost fresh (Keith, 1929).

The following comments on environments in which bones are found give some idea of the conditions of preservation to be expected.

Gravels

Bone preservation in gravels depends on the acidity and permeability, and on whether the deposit is anaerobic and waterlogged. Where undecalcified, gravels may lead to good preservation as in the Swanscombe fragments, but where decalcified by more acid conditions, the bones are poorly preserved (as in the Galley Hill remains). Waterlogged peat or peaty gravels and alluvial muds in Britain have produced many dark-brown stained and very well-preserved skulls, previously attributed to the 'River-bed' group (Huxley, 1862; Keith, 1929; Martin, 1935).

Chalk soil

Owing to its permeable nature, bones in chalky soil may be considerably eroded and fragile. Chalk takes a lot of washing off and even after the bone seems to be clean, it may dry white and continue to shed white dust. However, the washing must not be too vigorous, for if so the tartar deposits will be removed from the teeth. Chalk soil is usually soft but occasionally it may be marly and fairly consolidated, in which case special care and scraping with a penknife may be necessary.

Cave earth

The term cave earth has been used to cover a wide variation of deposits formed in caves, including consolidated dusts; clays, loams, sands and gravels laid down by water; and ungraded deposits consisting largely of talus mixed with clay. Cave deposits also include accumulations of limestone fragments cemented together by clay and calcium carbonate (breccias) or loosely aggregated and covered by films of precipitated carbonate (stalagmitic films). In the latter type of deposit, the bone may have been mineralized, or merely encrusted with the precipitated carbonates ('petrified'). It may be possible to chip away stalagmitic encrustations, but bones in the more solid cave breccias such as those containing much of the South African australopithecine material, require careful chipping and drilling or treatment in tanks of 10 per cent acetic acid to dissolve away the calcareous matrix. If sealed under layers of stalagmite, bones may be remarkably well preserved. Great care is needed in acid treatment of bone, and there is a danger of specimens being left too long. For preparatory work of this kind, Rixon (1976) should be consulted.

Clay

A clay matrix may lead to corrosion through acidity or, as in the Nottingham region where the subsoil is marl, it may ensure good preservation.

Sand

Sands, like gravels, vary considerably in their acidity or pH values. The sands at Mauer in Germany, where the fossil jaw of Heidelberg Man was discovered, and the gravels in the Barnfield pit at Swanscombe are calcareous and have a high pH so that the preservation of bone is good. The hot sands of both South America and Egypt have produced desiccatory conditions optimal for preservation of tissues such as skin and hair.

Salt soils

Occasionally, bones are recovered from very saline soils or marine shell-mounds, and then demand special treatment. Leechman (1931) suggests that such remains should be soaked in repeated baths of fresh water until no further salt efflorescences occur on drying. However, in a personal communication, Mr. B. Denston (1958) points out that every care should be taken while proceeding with such treatment, for if the bones are in a poor state of preservation, repeated washing may cause their disintegration. It therefore seems advisable to test one of the bones, and to note its preservation, before submitting all the remains to successive baths. If in a bad condition, it may be advisable to give only a preliminary soak before impregnating the bone with PVA (polyvinyl acetate) or an alternative hardening agent (e.g. Paraloid B72). This technique, as applied to the desalting of limestone ostraka from Egypt, has been described by Werner (1958).

10   Cleaning bones

Bones are not generally deceptive in regard to their strength, and it is usually evident on excavation whether they are very brittle and liable to break up easily when handled. However, if there is any uncertainty, one or two small post-cranial fragments can be submitted to a trial wash in warm water. If the fragments have stood up to this trial, then the rest of the skeleton can be washed in a similar way, particular care being taken with pieces that are very eroded. The water should not be very hot. No detergent is necessary and should never be used. Most osteological remains can be cleaned by using a knitting needle or hairpin for loosening the soil and then a nail- or tooth-brush with water. If there is earth filling broken long-bones, it should be loosened and shaken out as much as possible before soaking, as this will avoid extensive breakage of cancellous bone. Specimens must dry slowly, and not in hot sun.

If the bones are extremely brittle or squashed, the earth should be cleaned off with knife or pin and a soft brush, and a consolidant applied without further washing. This was necessary, for example, with a number of skulls excavated at Jericho, that suffered from extreme lateral compression. In Britain it is important to apply such procedures to all bones of pre-Roman age, as these are relatively scarce, but later material may also be treated thus if time allows.

Earth scraped from the surfaces of excavated bones should be saved as reference samples and for pollen analysis when this offers a possibility of dating evidence. Small fragments of bone considered to be of no use in reconstruction should be saved as samples for chemical analysis.

Few post-cranial bones need any particular washing procedure, but special care must be exercised while cleaning the skull. Before immersing in water, as much earth as possible should be removed from inside the brain-box; most can be carefully loosened with a knitting needle and removed through the foramen magnum. The ear hole (external auditory meatus) can be cleaned with a pin, but care should be taken to save any ear ossicles that may be in the soil. It should be remembered that in special circumstances even the brain or traces of it have been preserved within the skull, for example in bog burials, in desiccated bodies (mummies) and occasionally in burials in damp clay soils. Traces of brain preserved in adipocere were found in the skull of a Roman burial in Droitwich (Oakley & Powers, 1960). The facial region is the part of the skeleton most likely to break or collapse in washing. As much soil as possible should be removed, and even then, the combined weight of water and soil in the nasal cavity and maxillary sinuses may be sufficient to cause breakage. It is therefore important to support the skull amply while washing, in order to reduce this risk.

In the final cleaning of the nasal region it should be noted that the margin of the pyriform aperture, the conchae and the vomer are easily chipped or broken. The orbits, also, are formed of extremely thin bone that is easily fractured, particularly the lacrimals, which are paper-thin.

Spence and Tonkinson (1969) have discussed the value of an ultrasonic cleaning apparatus in separating adhering burial matrix from bones. This certainly is an excellent way of cleaning fairly strong bone, but the equipment is costly, and for the present will not be available to most archaeologists.

11   Use of consolidants

The topic of consolidating bone material is one that most archaeologists have had to consider at some time or other, and there is quite a deviation in opinion and procedure. However, as this is so important, it has been thought worth while to consider the relative merits of different materials and methods. Four common classes of 'preservative' are used, although only one really deserves this name.

The common preservatives

a) As Angel (1950) has already pointed out, hot *paraffin wax* can be very destructive. In some conditions, it penetrates the cracked bone and may split it further. Moreover, if this has been used as a preliminary field precaution, it may be found when attempting to dissolve it out in the laboratory that the bone is in a worse state of disintegration. If no alternatives are available, it is still really of doubtful value today, even though in the past it has been used on material from Ur (Woolley, 1949) and from Lachish (Risdon, 1939).

b) Although applied to many bones in the past, *shellac* is of doubtful use today. Only very occasionally does it penetrate and help to strengthen the inner areas of bone. Its application in a dilute form to the surface of a bone may assist temporarily to prevent flaking, but shellac-treated material in various museums in Britain clearly shows that after a few years the hard coat cracks and begins to peel away from the bone surface. This is especially so with bones from chalky soils where the surfaces are unevenly corroded. Moreover, as with wax, it is sometimes difficult to remove all the surplus solution, with the result that the thickness of the preservative may in some cases produce metrical inaccuracies.

c) Synthetic resins are the most useful consolidants. The most widely used in Britain have been Alvar 1570 and polyvinyl acetate dissolved in suitable media, the latter being preferable. Polyvinyl acetate should be dissolved in acetone or toluene. Alvar is now being replaced by Paraloid B72, a less flexible consolidant. When possible these synthetic resin solutions should be applied in the laboratory, not in the field. However, if the bones are extremely badly crushed and distorted, it may be advisable to apply a plastic solution after the preliminary cleaning in the field. A few fragments of bone should, of course, always remain untreated for the purpose of chemical analysis. Solvents for these consolidants are acetone and toluene (5–10 per cent).

In the laboratory, when possible, the bones should first be washed and then left to dry out completely. They should then receive a thorough impregnation with the dissolved synthetic resin. As it is not usually possible to allow the impregnation to take place under negative pressure (see below), sufficient time should be given for a thorough soaking. The bones should then be allowed to drain, so that the surplus will run off and leave a smooth surface. Alternatively, excess resin may be gently blotted off following the application of a little solvent to the surface. Unlike wax or shellac, resin consolidants help to retain the natural colour of the object, which may sometimes be important in sorting.

The application of consolidants is only necessary when the bone is liable to disintegrate in some way. This depends more on soil and climate than on chronological age. Even Palaeolithic fossils display great differences in toughness.

d) Fourthly, it is suggested that plaster of paris be used as little as possible. Although of value in supporting the soil or rock matrix around bones, its direct application to fragments may lead to difficulties in liberating them later in the laboratory.

e) For the consolidation of wet bones, PVA emulsions are sometimes useful. The commonest of these is an aqueous emulsion of polyvinyl acetate (Vinamul 6815), which can be thinned with water, and then applied by brush, spray or immersion under pressure (see below). In the case of all consolidation work, the bone should not be handled of course until absolutely dry.

12   Pressure impregnation

A method that can be of value in strengthening brittle bone, is the impregnation of a plastic consolidant under a negative pressure. It can be used in the laboratory or, with suitable adaptations, in the field. This was first employed in the field in the consolidation of the vertebrate bones from the waterlogged deposits of Mesolithic age at Star Carr, Yorkshire (Purves & Martin, 1950).

The equipment necessary for this procedure comprises the following:

a) Container of metal or glass large enough to hold a complete skull or long-bone. It should have an airtight cover and an air-outlet attachment.

b) Fairly strong suction pump for removing the air from the container.

c) Sufficient thick-walled tubing to attach to the pressure chamber and the pump.

d) Suitable bone consolidant (for example Paraloid B72 or polyvinyl acetate emulsion).

The procedure is to remove all the air from the container so that the synthetic resin will replace air in bone cavities. If loose bones are placed together in the container, they must be separated after impregnation, otherwise they will stick together. As most consolidants curdle if brought in contact with water, it is advisable to keep the specimens dry while under treatment. Excess consolidant solution or emulsion should be drained and wiped from all specimens, otherwise lumps of plastic will remain on the surface. For further discussion of this method, the reader is referred to Angel (1943).

One important warning is necessary. If the bone, or series of bones, appears to be particularly brittle, it is as well to test a small fragment of bone first before submitting the more valuable whole skulls or bones to the treatment. The reason for this is that pressure impregnation can occasionally lead to the collapse and disintegration of a specimen, as Denston (1958) found when treating a series of early Egyptian crania (personal communication).

13   Reconstruction of bones

If this has not already been done, the bone fragments should be separated, those of left and right bones being differentiated as far as possible. If consolidants have made any of the break-surfaces badly fitting, then the appropriate solvent should be used to clean them.

Every care should be exercised in reconstruction, for inferior work may result in considerable distortion. If a fragment is stuck badly, a solvent should be used to loosen it again, but it is advisable not to use force for this may result in fresh breaks.

A number of quick-drying cements are available. Balsa cement, Uhu or Durofix have been widely used. In the case of non-compact bone, it is advisable to apply the adhesive to both surfaces, and afterwards to wipe off the excess. PVA based adhesives can be used, but are not suitable for hot climates. Not more than two adjoining fragments should be stuck at the same time, but for quickness different areas of the same bone can be reconstructed to some extent independently.

Most post-cranial bones can be reconstructed on a flat surface, with supporting

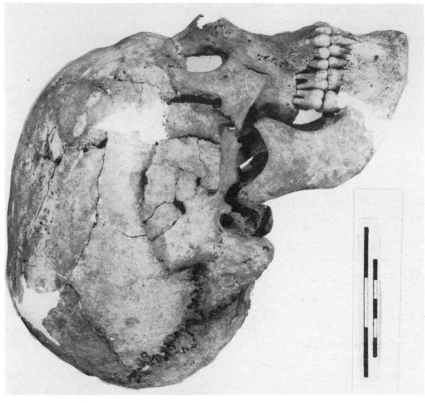

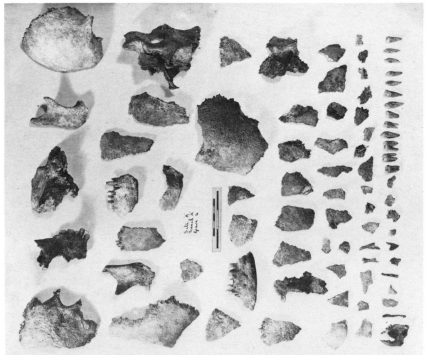

Plate 1.2   Skull of a man of the Dark Ages, from Cornwall: before and after reconstruction. By courtesy of the Duckworth Laboratory, Cambridge.

pillars of plasticine at strategic points if necessary. In the case of the long-bones, extra strength can be given to the shafts by placing a cylinder of wood in the central (medullary) cavity. The skull is best assembled in a sand box, where suitable depressions can be made as supports. Harrison (1953) found that dental wax and sellotape were useful in provisional reconstructions of very fragmentary skulls. Plasticine is also helpful in keeping bones in position while they are being stuck. Even when the skull is in more than fifty pieces, good reconstruction should be possible if the broken edges are clean and not eroded (as shown, for example, in Pl.1.2).

The skull bones should be separated before reconstruction is attempted. When possible, the vault and face are best reconstructed separately, and afterwards applied to one another. Before these two areas are completely fixed together, the condyles of the mandible should be applied to the glenoid cavities, checking to see that the teeth occlude properly. If the skull is extremely fragmentary, pieces of wood can be used as supporting props.

Unless the skull is very weak and fragmentary, no plaster of paris should be used to fill in gaps. If this is unavoidable, then the preparator should avoid covering any craniometric points, and as much of the endocranial (internal) surface as possible should be left visible. Plastered surfaces should be made smooth. It is better not to colour it to match the bones, as the natural white clearly shows future workers what is false.

The mandible should not be glued or plastered to the vault.

Finally, both during and after the reconstruction of cranial remains, correct orientation of the pieces should be attempted. Incorrect orientation may lead to the distortion of vault form, and may also give a wrong impression as to the morphology of a particular area. In Figure 1.1, for example, it will be seen that if the front of the vault is not correctly placed, the size of the brow ridges and degree of frontal recession may be unduly accentuated.

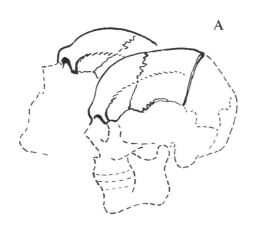
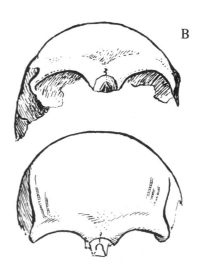

Figure 1.1   The result of good and bad orientation of a bone fragment, in this case from the skull vault. In both the lateral (A) and facial (B) aspects, poor orientation as shown in both upper drawings produces a more 'primitive' appearance than is in fact the case.

14   Burnt bones

1.   Cremations

In recent years it has become increasingly evident that cremated remains can yield data of immediate and comparative value to the archaeologist. A number of authors, particularly Gejvall (1947, 1951, 1969), Gejvall & Sahlström (1948), Lisowski (1955, 1956), Weiner (1951), Wells (1960), Spence (1967), Merbs (1967), have made reporting on cremations a more disciplined procedure. Van Vark (1974) has even demonstrated that multivariate statistical methods may be applied to the analysis of such remains. Sometimes, of course, the pieces will yield little information, and in the Ingelby Pagan-Danish remains, for example, the burning and subsequent deliberate pounding was so thorough that hardly any identification and description was possible (Cave, 1956). There is no doubt, however, that eventually cremations may prove quite as interesting for comparing cultures as metal or ceramic finds, and there is already some evidence to show that cinerary fragments may differ in certain ways between groups. It is evident that a study of such material is as important as that of the urn or the strata containing it.

As in the description of unburnt material, it is urged that the report should be as complete in itself as possible. First, the environment should be noted, as this may possibly provide a basis of classification at some future date. Burnt human bones may be found:

a) In or by the side of cinerary urns (upright or inverted).

b) Scattered in ash or ordinary soil.

c) Mixed with fragments of animal bone (which may also be cremated, as Lisowski showed in 1956).

d) Mixed with unburnt human bone, either larger or broken into fragments of similar size.

e) In burial mounds, open fields or domestic hearths.

Before further features can be reported, a preliminary sorting and cleaning will be necessary. It is certainly advisable to separate all the larger fragments from the small ones, and if time allows the remains could be gently washed on a sieve of 2 mm mesh, and left to dry on trays covered by newspaper. The remains of teeth should then be removed and the fragments of bone shaken in a 5 mm sieve in order to remove any remaining earth and bone dust (Lisowski, 1956). Larger stones and charcoal fragments can be removed by hand, the latter being saved as a possible source of radiocarbon dating.

It will now be possible to discuss a number of general features including:

a) Average size of the fragments (estimated in millimetres).

b) Preservation (related to the soil).

c) Colour (urn remains are usually lighter).

d) Texture of the bone in general (e.g., slightly charred or heavily calcined).

e) Degree of brittleness, fissuring, distortion and twisting (Pl. 1.3). This twisting will give some idea of the heat of the funeral pyre, for as Angel & Coon (1954) pointed out, the smaller amounts of heat, which produce only charring, or whitening through removal of organic matter, scarcely distort the bone at all.

f) Overall weight of the remains.

g) Any evidence to show that the bones were deliberately broken after burning (the size of the fragments is an important factor).

Although there is no doubt that shrinkage is liable to take place in certain bones, its detec-

Plate 1.3   Fragments of cremated bone from an Irish cairn (slightly reduced). By courtesy of the National Museum of Ireland.

tion and the degree are usually not ascertainable. It is also uncertain whether the fissures, particularly of the parallel type, are at all related to the anti-stress lines revealed by the split-line technique (Benninghoff, 1925; Tappen, 1955; Evans & Goff, 1957).

The sorting of burnt bones into anatomical groups, such as parts of skull, vertebral column, ribs, unidentified long-bones, tarsals, phalanges, metacarpals, metatarsals and more doubtful fragments, is a somewhat specialized procedure which may be too difficult for the excavator. If this proves to be possible, however, then the weights of the various groups (in grams) can be recorded, and their percentages of the total number of burnt fragments, and of the identified remains. If complete sorting is impossible, it is as well to note the identifications of as many of the burnt bones as possible, for these

may still suffice to show whether or not all parts of the body were represented. The following fragments particularly should be looked for, as they provide evidence of the sex, the age and the number of individuals present:

a) Fragments of femur shaft (especially in the region of the linea aspera), femur head, mastoid processes, external occipital protuberance, and superciliary ridges, which by their size, thickness or ridging may suggest the sex of the individual.

b) The presence of more than one chin region of the mandible, odontoid process of the axis or more than one atlas vertebra or sacrum, will show the presence of at least two individuals. A similar conclusion is necessary if there are more than two supraorbital margins, glenoid fossae, condyloid processes of the mandible, humerus heads, individual tarsal bones, or petrous portions of the temporals. Skull (vault) fragments of noticeably different thicknesses may also show the presence of more than one person.

c) Estimates of age depend upon finding such fragments as teeth; proximal and distal parts of long-bones, vertebrae, and other bones displaying age indications; and skull vault (see also under 'Estimation of age').

The teeth may be fairly complete if the body was not submitted to great heat, but sometimes the temperature was enough to splinter off all the enamel, leaving only a core of dentine intact. In whichever condition they are, it should still be possible to note traces of cementum-apposition on the root, secondary dentine in the pulp cavity and closure of the root orifice. The degree of dental attrition (wear) may be another useful indicator of age, and even where the enamel is missing, it may still be possible to detect wear on the 'occlusal' tips of the dentine.

Any areas of bone with sutural junctions that do not unite until a certain age are worth particular note, and may permit an age estimation either with or without consideration of the size factor.

Skull thickness in immature individuals bears a rough relation to age, and it is suggested that possible children's skull bones should be compared with the average dimensions worked out by Roche (1953).

Sometimes it is possible to note anomalies and evidence of disease in cremated material, although this is by no means usual. Lisowski (1956), for example, notes a sutural bone in one group of remains. The larger caries cavities in the teeth may be found, even if dentine alone remains. Osteoporitic pitting on cranial fragments and rheumatic changes, particularly on vertebral fragments, are also sometimes distinguishable. It would be extremely difficult, however, to recognize pre-mortem sword or axe cuts, for burnt fragments often show 'clean' breaks which could easily be mistaken for battle injuries.

2. Accidental exposure to fire

Sometimes only a small portion of a skeleton shows exposure to fire. In most cases these are no doubt accidental, although intentional mutilation cannot always be ruled out completely. The results of the heat in this case are usually restricted to charring (blackening) of the bones. Whether the heat was applied before or after the body was reduced to a skeleton, is a problem which cannot yet be answered satisfactorily. Deeply blackened bone may suggest that flesh was still present, and Camps (1953) has pointed out that in certain recent material, the 'black coloured glaze' suggests that burning occurred whilst blood was still present. Another point which can be used to ascertain whether a bone or even a complete skeleton was already dry when exposed to heat, is that if it were so, fissuring and twisting would probably be negligible, and there would be less sign of calcination.

15   Mixed bones

It sometimes happens that the excavator uncovers remains of two or more individuals that are thoroughly mixed. Unless the bones are extremely fragmentary and eroded, it is usually possible to separate them, at least to some extent. Post-cranial material presents the major problem of segregation, especially if the bones are broken and show corrosion. In all cases, the following lines of examination should enable some division to be made.

a) *Variation in colour*

In some cases, bones vary in colour from one region of the skeleton to another, and this criterion is therefore best used in conjunction with others.

b) *Degree of preservation*

This cannot in itself be conclusive, for in some specimens certain areas may disintegrate before others.

c) *Size differences*

Noticeable size variations between similar bones are usually due to their belonging to different individuals rather than to gross asymmetry. The bones especially concerned here are the long-bones, talus, calcaneus, metatarsals, metacarpals, scapula, clavicle, pelvis and patella.

d) *Bone shape*

The general form as well as the prominences for muscle insertion, often clearly distinguish two like fragments.

e) *Articulation*

It may be found that parts of certain bones articulate particularly well, suggesting that they belong to the same individual. The bone areas especially concerned are the femoral head and acetabulum; the femoral condyles and proximal end of the tibia; the proximal part of the ulna and the distal end of the humerus; and the vertebrae.

f) *Anomalies*

Evidence of a metopic (inter-frontal) suture may be found on more than one skull fragment. Osteo-arthritic lipping, especially on vertebrae, may help to separate young and middle-aged adults. Injury, such as a sword cut, may extend over more than one fragment. This latter fact was an important point in associating fragments of a Saxon skull that was excavated on Portsdown Hill in Hampshire.

g) *Wear of the teeth*

Differences in attrition on similar teeth can be of particular value in separating adult jaws.

h) *Occlusion*

Testing their interlocking fit or occlusion is an important way of associating mandible and upper jaw – especially if both are fairly complete.

i) *Sex differences*

There may be features of the pelvis, skull and other bones that enable them to be associated. For example, large brow ridges are usually correlated with large mastoid processes. Prominent nuchal ridges and external occipital protuberances are more likely to be associated with a large masculine than with a gracile feminine mandible.

j) *Sutural variation*

Suture margins from the same area of skull, showing differences in the degree of sutural obliteration, may help in the segregation of bones representing several crania. On the other hand similar types of suture may not only help to associate fragments, but the pieces may be found to fit.

If only a few individuals are represented in a mixed assemblage of bones, it may be possible to suggest by such factors as immaturity or skeletal ruggedness, which skull may belong to a particular set of post-cranial remains. The atlas vertebra, when present, is important in associating the skull with the vertebral column, although occasionally more than one atlas may at first appear to articulate with a given pair of occipital condyles.

If bones have been numbered according to depth excavated area before sorting, it should be possible to refer them to original positions.

16   The nature of bone

To the excavator, bone seems to be merely the solid and unchanging part of the human organism, but during life it was in fact just as living as any other tissue of the body. Not only does it grow and suffer from disease, but it has the capacity to heal after an infection or breakage. It is also sensitive to the needs of the individual, so that unnecessary bone is resorbed, as with the tooth-socket after an extraction, the end of a bone after amputation, or a paralysed limb when the muscles cease to demand a strong skeletal structure. Conversely, bones will become thicker and stronger if greater demands are made upon them. As a result of growth studies, we now know that even during the adult age period varying amounts of bone expansion occurs in different regions of the skeleton (Kendrick & Risinger, 1967; Garn, Wagner, Rohmann & Ascoli, 1968).

Another wrong impression is that bone is purely compact inorganic matter rather like chalk, whereas in fact, it has an organic framework of fibres and cells amongst which the inorganic salts, notably calcium phosphate, are deposited in a characteristic fashion. This combination of about one-third organic matter and two-thirds inorganic matter gives the bone both resilience and toughness, as well as hardness and rigidity (Fig. 1.2). It is a simple matter to demonstrate the presence of organic constituents by placing a bone for a while in an acid. This removes the salts and leaves the organic part of the bone, mainly as collagen bundles, which still retains the shape of the untreated specimen, and in such a condition it is extremely flexible. Thus, a decalcified long-bone can easily be tied in a knot (Fig. 1.3b), tending to go back to its normal shape when

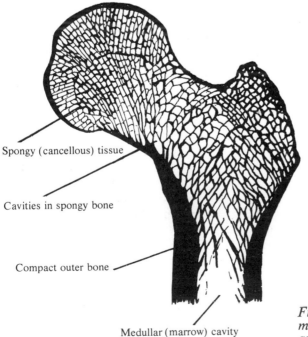

Spongy (cancellous) tissue

Cavities in spongy bone

Compact outer bone

Medullar (marrow) cavity

Figure 1.2   Internal structure of a bone as revealed in a section of a femur (proximal end).

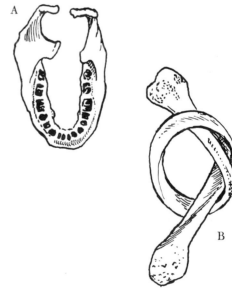

Figure 1.3   A   Severe lateral compression in a mandible that has been partially decalcified in an Irish bog.
B   Modern human fibula that has been rendered completely pliable after decalcification in an acid.

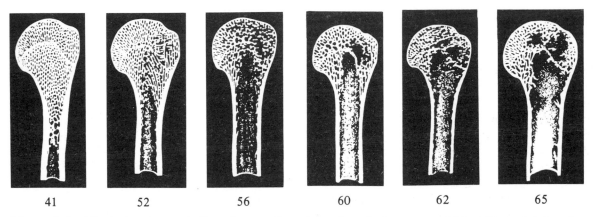

41   52   56   60   62   65

Figure 1.4   Variation in cortical and cancellous bone in relation to age (After Nemeskéri et al., 1960).

untied. Deformity of a milder nature easily occurs if the calcium content of buried bone is considerably reduced. Such a case of distortion is seen in a mandible from an Irish peat bog (Fig. 1.3a). The rapidity with which a bone loses its organic content depends very much upon climate and soil. In Britain reduction of this organic matter to a negligible amount may take many centuries. Even Iron Age and Saxon bones may, in the laboratory on decalcification, show sufficient organic residue to maintain at least their shape, if not their size. Smaller quantities of organic matter may remain for many thousands of years and this fact has been exploited in the relative dating method by nitrogen content (Oakley, 1958).

It would appear that in the complex chemical process of fossilization, microscopic sections of bone may reveal at such high magnifications as × 20 000 apparent evidence of collagen fibres, even though the organic content of the bone has been mainly replaced (Race, Fry, Matthews, Wagner, Martin & Lynn, 1968).

Considered macroscopically (that is, viewed by the naked eye) there are two forms of bony tissue:

a) Spongy or cancellous tissue

This comprises much of the interior of a bone. The spongework is composed of lamellae or plates that are so arranged in the bone as to be of maximum value in withstanding pressure and tension.

b) Compact or dense tissue

This forms a complete outer casing of the bones. Considerable variation occurs in the thickness of this cortical bone, both in relation to the sexes and different age groups (Virtama & Helelä, 1969; *see* Fig. 1.4).

II Description and study of human bones

1 The bones of the skeleton

In this short handbook detailed anatomical descriptions would be out of place, especially as most libraries are equipped with such textbooks. A short classification, however, may be useful in field work and in the general handling of bones. When possible, first-hand acquaintance with bones is advisable, and it is more important to get a visual picture of a bone than to attempt to memorize elaborate descriptions. A good knowledge of the general morphology of the human skeleton is certainly necessary in the recognition of fragmentary remains, particularly when one cannot rely on such features as bone size or the presence of specific distinguishing points.

The following is a brief summary of the bones of the skeleton and their relationship to one another. The scheme is similar to that given by Furneaux (1895).

The skeleton (Figs 2.1–2.11)
skull (Figs 2.2 and 2.3)
vertebral column (Figs 2.4 and 2.5)
ribs and breastbone (sternum) (Fig. 2.5)
shoulder girdle (Figs 2.6)
upper limbs (Figs 2.7 and 2.8)
hip (pelvic) girdle (Fig. 2.9)
lower limbs (Figs 2.10 and 2.11)

These major areas may be subdivided to give specific bones, and any character that particularly distinguishes them has been noted. The number of each type of bone is also given, although numerical variations can occur (see Schultz, 1930).

With this basic skeletal knowledge, it is now possible to identify human remains, mixed or as well defined inhumations. At this stage, one should be able to mark (or fill in) the relevant identified skeletal elements on standard checking sheets of the kind seen in Figs 2.12–2.14.

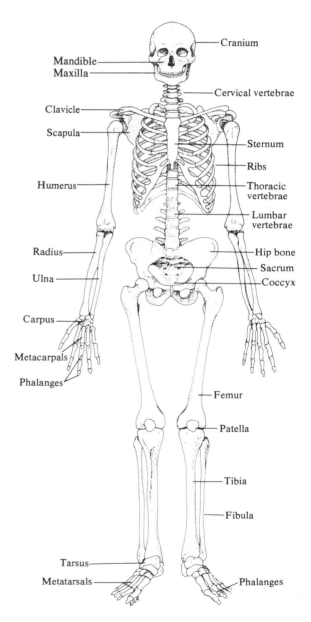

Figure 2.1 The human skeleton.

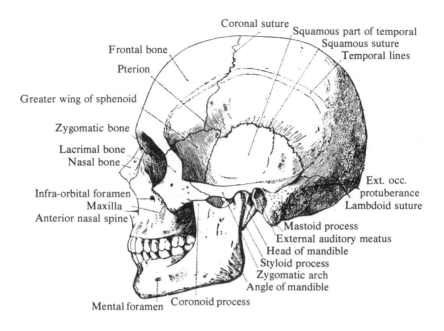

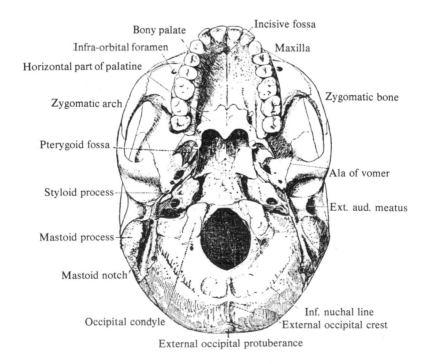

Figure 2.2   Left lateral and basal aspects of the skull.

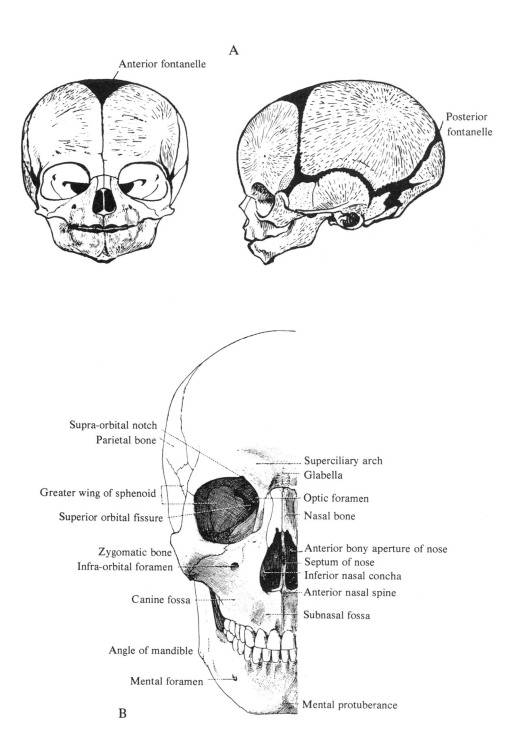

Figure 2.3 A, Facial and lateral aspects of the skull of a newborn. B, Facial anatomy of an adult skull.

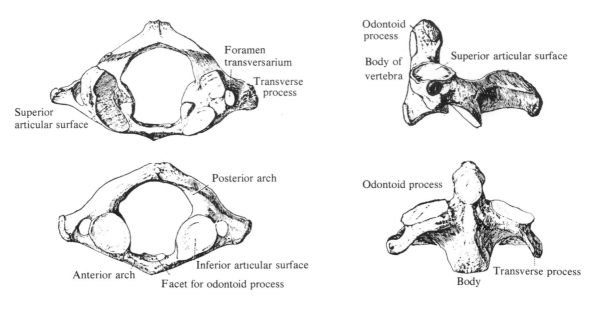

Foramen
transversarium

Transverse
process

Superior
articular surface

Posterior arch

Anterior arch

Inferior articular surface

Facet for odontoid process

The atlas (cervical) vertebra, upper surface (top),
lower surface (below)

Odontoid
process

Body of
vertebra

Superior articular surface

Odontoid process

Transverse process

Body

The axis (cervical) vertebra, left side (top),
front view (below)

Superior articular process

Spine

Facet for
tubercle of rib

Transverse process

Body of vertebra

Facet for tubercle of rib

Superior articular process

Facet for tubercle of rib

Spine

A thoracic vertebra, upper surface (top), left
side (below)

Spine

Superior articular
process

Foramen
transversarium

Spine

Inferior articular process

Foramen transversarium

A lower cervical vertebra, upper surface (top), lower
surface (below)

Figure 2.4   Characteristics of cervical and thoracic vertebrae.

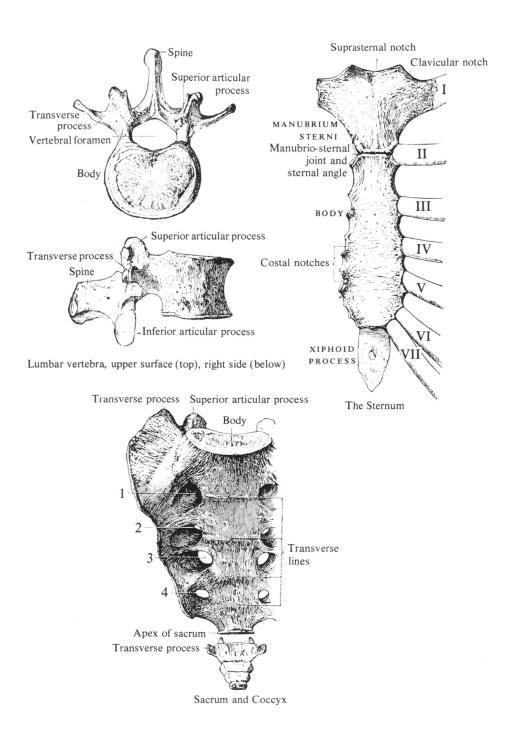

Figure 2.5   *Characteristics of the sternum and lower part of the vertebral column.*

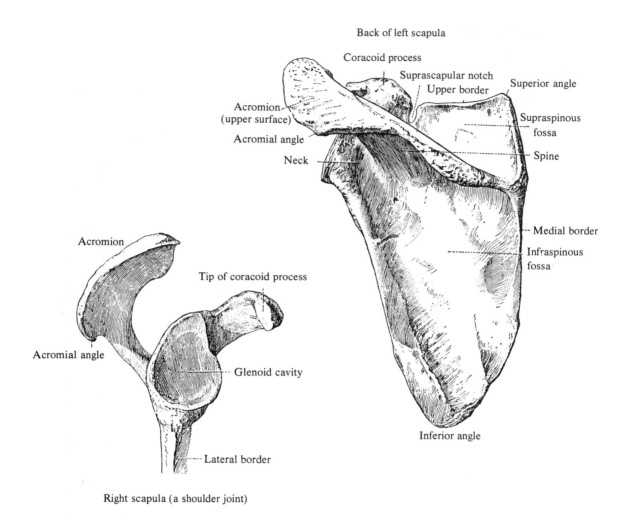

Back of left scapula
Coracoid process
Suprascapular notch
Upper border
Superior angle
Acromion
(upper surface)
Supraspinous
fossa
Acromial angle
Spine
Neck
Medial border
Infraspinous
fossa

Acromion
Tip of coracoid process

Acromial angle
Glenoid cavity

Lateral border

Inferior angle

Right scapula (a shoulder joint)

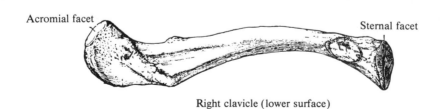

Acromial facet
Sternal facet

Right clavicle (lower surface)

Figure 2.6   Bones of the shoulder girdle.

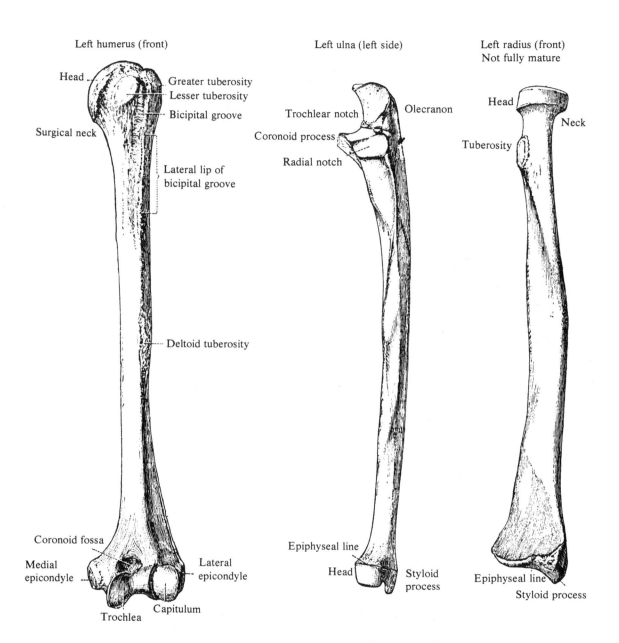

Figure 2.7 Bones of the arm.

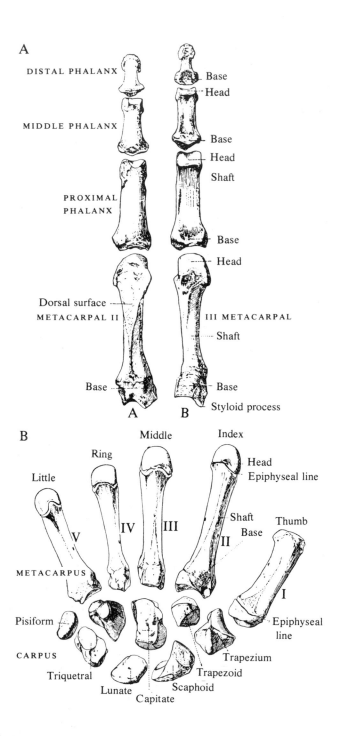

Figure 2.8   Bones of the hand. A, digit II (dorsal surface) and III (anterior-palmar surface). B, all the carpal and metacarpal bones (palmar aspect).

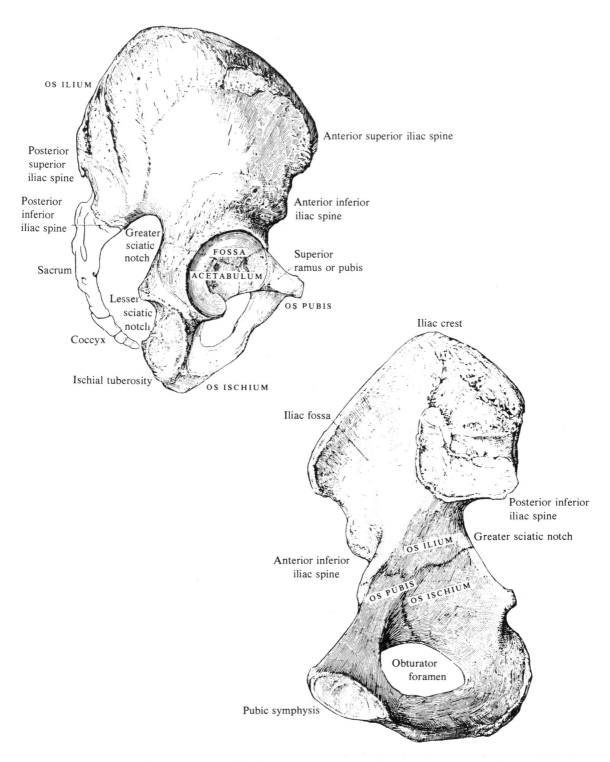

OS ILIUM

Posterior
superior
iliac spine

Posterior
inferior
iliac spine

Sacrum

Greater
sciatic
notch

Lesser
sciatic
notch

Coccyx

Ischial tuberosity

Anterior superior iliac spine

Anterior inferior
iliac spine

FOSSA
ACETABULUM

Superior
ramus or pubis

OS PUBIS

OS ISCHIUM

Iliac crest

Iliac fossa

Posterior inferior
iliac spine

Greater sciatic notch

OS ILIUM

Anterior inferior
iliac spine

OS PUBIS

OS ISCHIUM

Obturator
foramen

Pubic symphysis

Figure 2.9   Outer and inner views of the right hip bone.

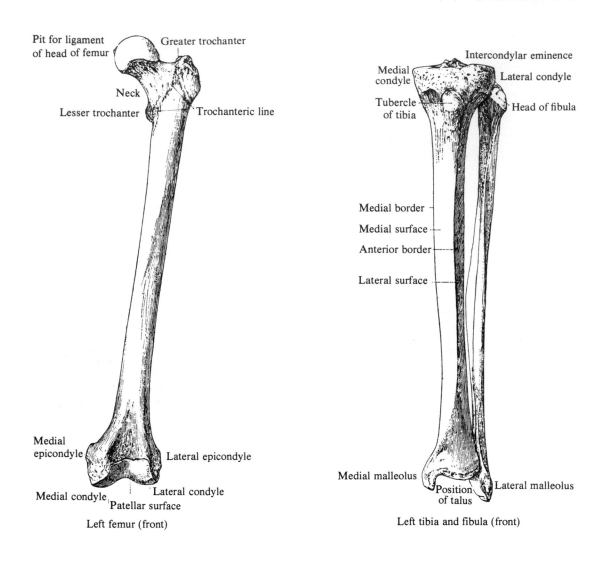

Pit for ligament
of head of femur

Greater trochanter

Neck

Lesser trochanter

Trochanteric line

Medial
epicondyle

Lateral epicondyle

Medial condyle

Lateral condyle

Patellar surface

Left femur (front)

Medial
condyle

Intercondylar eminence

Lateral condyle

Tubercle
of tibia

Head of fibula

Medial border

Medial surface

Anterior border

Lateral surface

Medial malleolus

Position
of talus

Lateral malleolus

Left tibia and fibula (front)

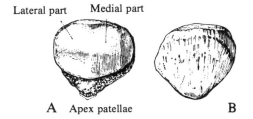

Lateral part   Medial part

A   Apex patellae   B

Left patella. (A) back, (B) front

Figure 2.10   Bones of the leg.

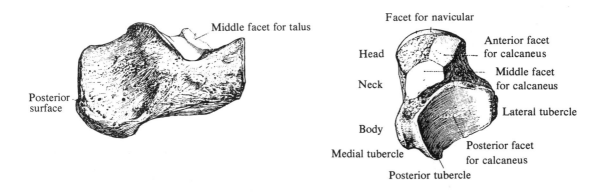

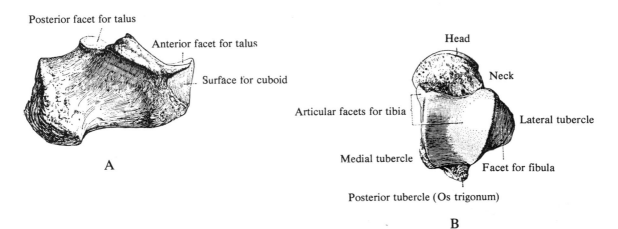

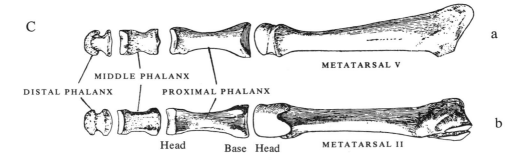

Figure 2.11   Bones of the foot: the left calcaneus and talus. A, lateral (top) and medial (bottom) aspects of the calcaneus. B, lower (top) and upper (bottom) aspects of the talus. C, metatarsals and phalanges. a, upper surface. b, lower surface.

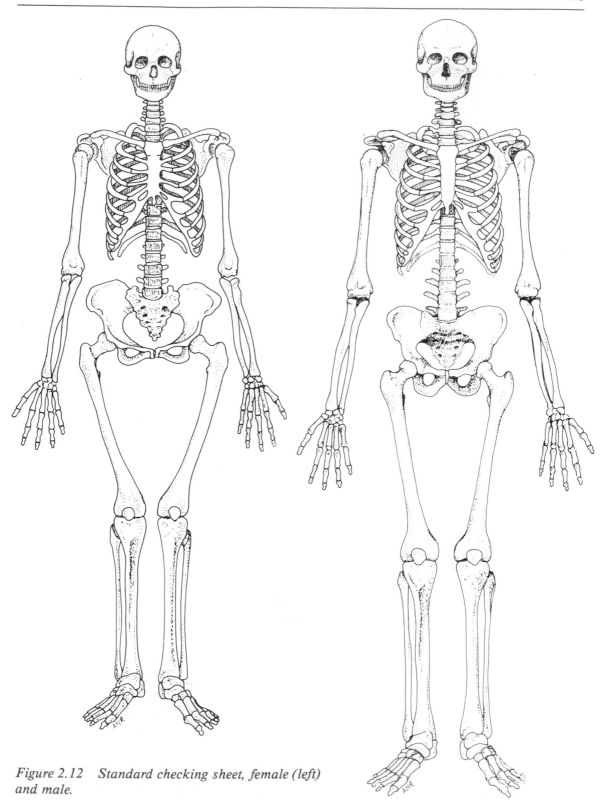

Figure 2.12 Standard checking sheet, female (left) and male.

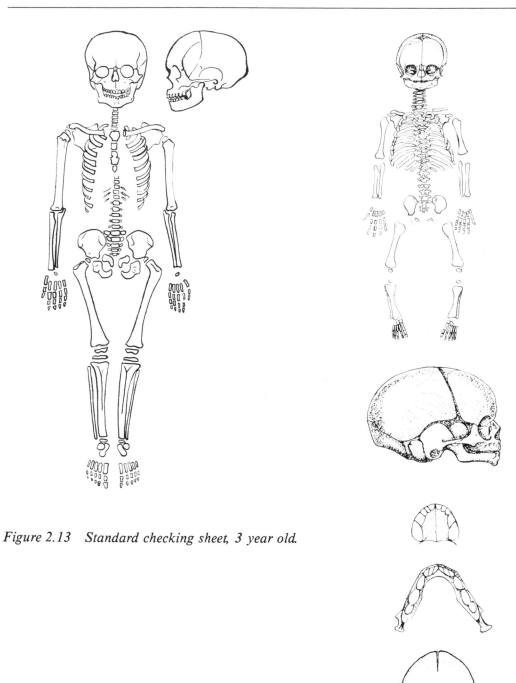

Figure 2.13 Standard checking sheet, 3 year old.

Figure 2.14 Standard checking sheet, embryo.

Skull

Cranium	*Occipital (1)* with nuchal crests, external occipital protuberance, and foramen magnum.
Calvarium (brain-case)	*Parietal (2)* with parietal foramen; temporal lines continued from frontal.
	*Frontal (1) the supraorbital ridges are important in sexing.
	Temporal (2) has squamous, mastoid, petrous and tympanic parts.
	*Sphenoid (1) irregular, with greater and lesser wings; including the sella turcica which houses the pituitary gland.
	Ethmoid (1) a cuboidal, sieve-like bone.
Face	**Malar (2)* zygomatic or cheek bone.
	*Maxilla (2) with the maxillary sinus and alveolar process.
	Palatine (2) back part of the palate.
	Nasal (2) forms the 'bridge of the nose'.
	*Lacrimal (2) a very thin, scale-like bone.
	Inferior nasal concha (turbinate) (2) within the nasal cavity; scroll-shaped.
	Vomer (1) separates the left and right nasal areas.
Mandible	Comprises two symmetrical halves. Consists of a body *(corpus mandibulae)* and two *rami*. The body has a median symphysis and a pair of mental foramina. The rami branch into the condyles and coronoid processes.
Additional Bones	*Sutural (Wormian) (1 to over 20).* Along the cranial sutures.
	Hyoid (1) supports the tongue. Body plus greater and lesser horns. (Below this, the thyroid cartilage may be found calcified.)

Vertebral column

Cervical (neck) (7)	Small body, with spinous and lateral processes. Has the characteristic foramen transversarium. The 1st (Atlas) and 2nd (Axis) cervical vertebrae are specialized.
Dorsal (thoracic) (12)	Have facets for articulation with ribs. The body (centrum) and transverse (lateral) processes are larger.
Lumbar (5)	Very large bodies, with long spinous and transverse processes.
Sacral (5)	Commonly united to form the sacrum.
Coccygeal (4 or 5)	Degenerate. These bones (forming the coccyx) continue the curve of the sacrum.

Ribs

True ribs (7 pairs)	Connected by their own cartilage with the breast-bone (sternum).
False ribs (5 pairs)	8th, 9th and 10th pairs are connected with the cartilage above it. 11th and 12th pairs are floating, being fixed at only one end.

*Bones articulating to form the orbit.

Sternum (breast-bone)

Manubrium (1)	Irregularly quadrilateral in shape.
Body (1)	Long and blade-like. Displays a series of lateral facets for the costal (rib) cartilages.
Xiphoid process (1)	Variable in shape. Begins as cartilage, but usually becomes incompletely ossified in the adult.

Shoulder-girdle

Clavicle (collar-bone) (2)	's' shaped. Articulates with sternum and scapula.
Scapula (shoulder-blade) (2)	Consists of a triangular body, the lateral angle displaying the glenoid cavity. Also a coracoid process and spine.

Upper limbs

Humerus (2)	With a long shaft, a proximal head, and distally situated articulatory area.
Forearm	*Ulna (2).* With large proximal head and slender shaft. *Radius (2).* The upper head is flat and circular, the distal end being much larger.
Hand	*Carpus (wrist). Carpal bones (8 each hand).* Arranged in two rows. Comprise the scaphoid, lunate, triquetral, pisiform, trapezium, trapezoid, capitate and hamate. *Metacarpal bones (5 each hand).* Excepting those of the thumb, they are all very similar. *Phalanges (14 in each hand).* There are 2 in the thumb and 3 in each finger.

Pelvis (Ossa innominata (2))

Ilium (2)	The flat semi-circular blade above the depression for the femoral head.
Ischium (2)	Below the ilium, it is narrow and thicker. Forms part of the margin for the large obturator foramen.
Pubis (2)	Flattened and less massive than the ischium. Forms the rest of the obturator foramen margin.
Acetabulum	Not a bone but a hemispherical depression which articulates with the head of the femur. All three divisions of the hip bone take part in its formation.

Lower limb

Femur (thigh-bone) (2)	Longest bone of the skeleton. The proximal head is connected with the cylindrical shaft by the neck. Distally, there are condyles which coalesce in front.
Patella (knee-cap) (2)	Roughly heart-shaped, appearing as a thickened disc.
Lower leg	*Tibia (2).* The large upper extremity articulates with the femur. *Fibula (2).* Slender throughout its length.
Ankle—tarsal bones (7 each limb)	Have an irregular cuboidal form. Consist of the calcaneus, talus, cuboid, navicular, and 3 cuneiform bones.
Foot	*Metatarsal bones (5 each foot).* Excepting those of the big-toe, they closely resemble one another. *Phalanges (14 in each foot).* Have a similar arrangement to that found in the hand.

2 The differentiation of human from other larger mammal bones

Even in the most experienced hands, it is not always easy to identify human remains. Fifty years ago a possible new skull of an early hominid was discovered in Java but turned out to be the proximal epiphysis of an immature fossil elephant humerus (Fig. 2.15)! Similarly, from East Africa, a suspected early hominid skull turned out to be a fragment of a flat bone plate of a turtle (Fig. 2.15). Of more recent date, it can be recounted that on more than one occasion carnivore phalanges (especially bear) have been mistaken for human finger bones.

Because the differentiation of human from other bones is an important part of preliminary sorting work, it would seem useful to make some brief comment about the sort of variation that occurs in mammals, and some areas which may produce special problems. Difficulties in sorting of course depend upon the degree of fragmentation and on the nature of the mixing – how many mammal bones in the human size range are together. Problems may increase with the numbers of species present. Bone samples from more recent sites, where mixing may only be with restricted numbers of domestic species, are likely to present the least problems.

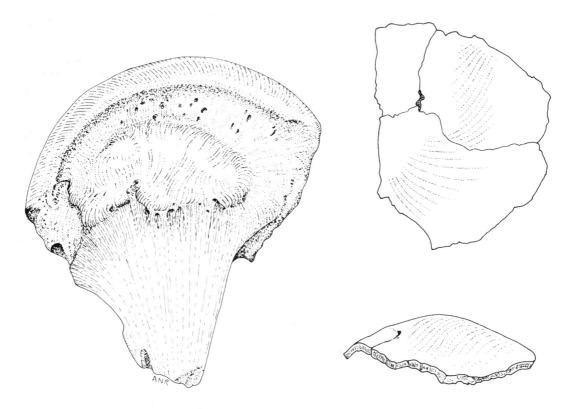

Figure 2.15a Proximal epiphysis of an immature fossil elephant humerus confused for an early hominid skull.

Figure 2.15b Flat bone plate of a turtle confused for a hominid skull.

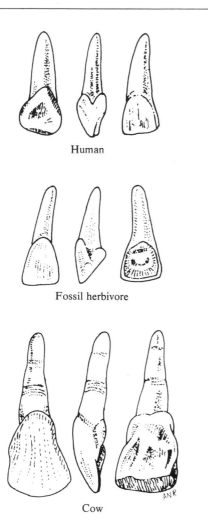

Human

Fossil herbivore

Cow

Figure 2.16   Human and herbivore incisors.

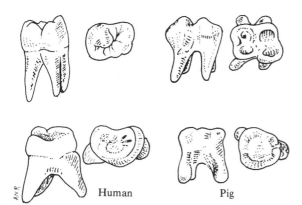

Human   Pig

Figure 2.17   Pig and human molar teeth.

a) The skull

Cranial bones are highly variable in shape in mammals, and generally present no great problems in terms of separating off human fragments. Our large brain-case and distinctive small face usually permits easy identification, but at times small fragments from some areas may be misleading. For instance, fragments of nasal bone from a dog-sized mammal may look like pieces of human nasal, and similarly parts of the basi-occipital area or malar bone might be confused. Small pieces of mandible of a sheep/deer-sized mammal, particularly a fragment of condyle, coronoid process or even part of the lower border of the mandibular body, could be confusing.

In the case of some immature mammals (e.g. horse, sheep) fragments of brain case could be mistaken for an immature human, although a closer study of the suture form and the nature of the endocranial surface should help to sort this question out.

Teeth are to some extent distinctive, although fractured pieces or heavily worn fragments may present problems. The type of problem specimen that can occur is seen in Fig. 2.16, which is not a human incisor but that of a fossil herbivore. One is reminded here of the claimed fossil primate *Hesperopithecus* which turned out to be a peccary, and of the early hominid molars from Trinil, Java, which are fossil orang. In more recent samples, confusion can occur between worn pig and human posterior teeth (Fig. 2.17).

b) The vertebrae

Human vertebrae tend to be fairly distinctive, and are somewhat modified as part of our evolution to bipedalism. The vertebral body tends to be short in the head-to-sacrum direction but broad laterally, compared with other mammals of comparable size. Also, the intervertebral surface of these bodies tends to be rather flat, with sharply defined margins.

The rounded nature of the vertebral body in man is rather distinctive in land mammals, but it should be noted that in some marine mammals, there is even more perfect rounding. Fragments of neural spine from such medium-size mammals as bear and goat could be mistaken for human specimens.

Some misinterpretations can of course be the result of just insufficient knowledge of comparative anatomy. For example, an axis of a wild pig from Ghardalam in Malta was mistakenly identified as human and accepted as such for a generation or so (Fig. 2.18).

c) The scapula

Although this bone is a fairly complex shape, and there is much variation in mammals, fragments may cause problems. The shape of the main blade is distinctive (short and triangular) in man, but small fragments of the thin bone may not be easy to identify with certainty. The articular area in man merges immediately into the blade, but most mammals of comparable size show some narrowing into a neck region beyond the articular area. The elliptical glenoid cavity would need careful identification if in a fragmentary state.

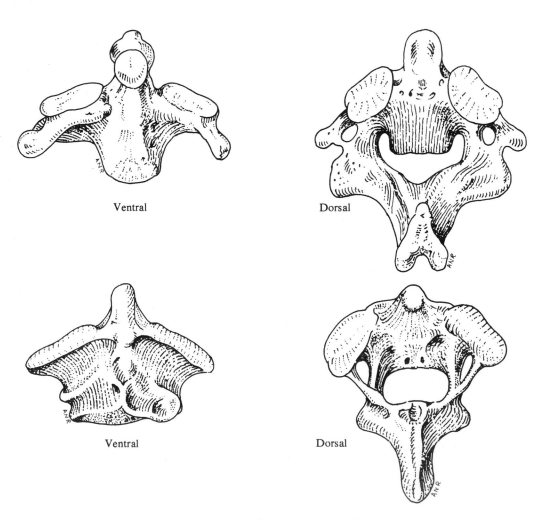

Ventral   Dorsal

Ventral   Dorsal

Figure 2.18   Axis vertebrae of a human (above) and wild pig (below).

d) The pelvis

The human pelvis has a highly specialized form, adapted for bipedal locomotion, and except in very small pieces, would not normally be mistaken for any other. For example, unlike mammals in general, the iliac blade is S-shaped, short and broad (antero-posteriorly). However, the acetabulum region may be something of an exception if only fragmentary pieces are available. In fact even here, small-scale differences in the shape of the articular region, depth of the acetabulum, and general shape and thickness of surrounding bone surfaces, should allow identification without too much trouble.

e) The long bones

The various long bones of the limbs need only brief consideration, and to a great extent can be separated on shape and size. Smaller fragments of the shaft of larger long bones can however look quite similar to human ones, although in man the shafts tend to be straighter and have a less rugged appearance compared to most mammals of a similar size. There are also differences in the thickness of the cortical tissue in different mammals (as well as between different types of bone). Parts of some articular ends can be deceptive. Fig. 2.19 gives examples of the proximal ends of a range of femora and humeri. In the case of

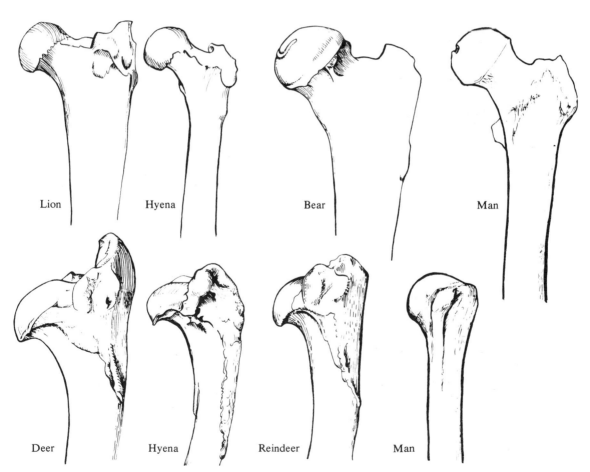

Figure 2.19   Proximal ends of femora (top) and humeri.

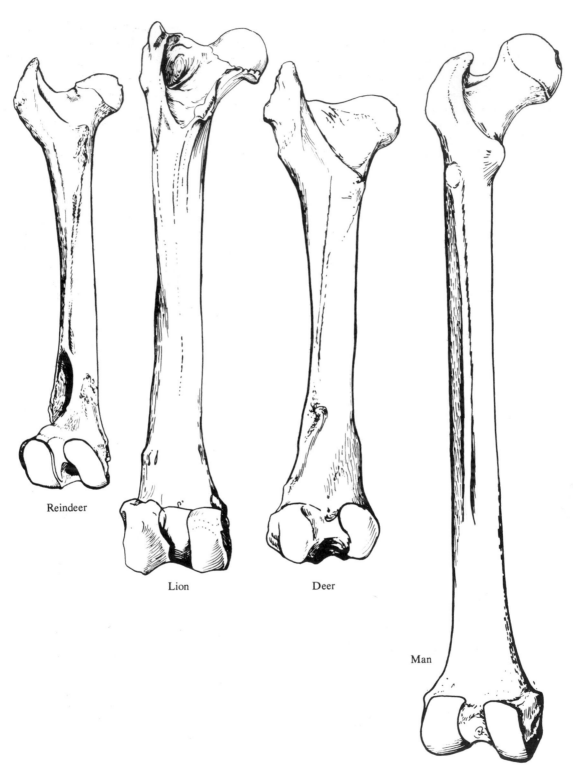

Figure 2.20   Femora from a range of mammals.

the femur heads, they look roughly similar, although in fact there are small size and shape differences; also the femoral necks tend to be shorter than in man and there is variation in the narrowing of the shaft distally.

In the case of the humerus, the tuberosities tend to be higher than the humerus heads other than in man, and the shaft is bowed more. Fig. 2.20 shows a few views of the whole femur in man compared with carnivore and herbivore representatives. It can be seen that the lion femur in this small group is most similar to man, though on closer viewing, one can see that femoral head shape is slightly different, the trochanters are smaller, and there is no marked linea aspera down the back of the shaft. Moreover, the distal condyles show small-scale but definite shape

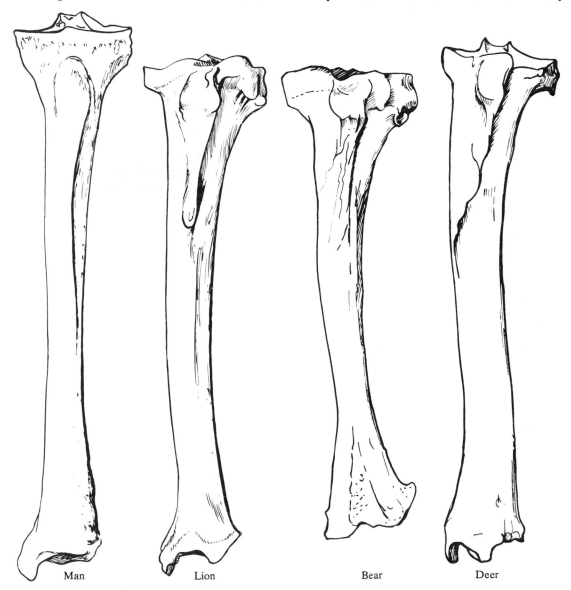

Man   Lion   Bear   Deer

Figure 2.21   Mammalian tibiae.

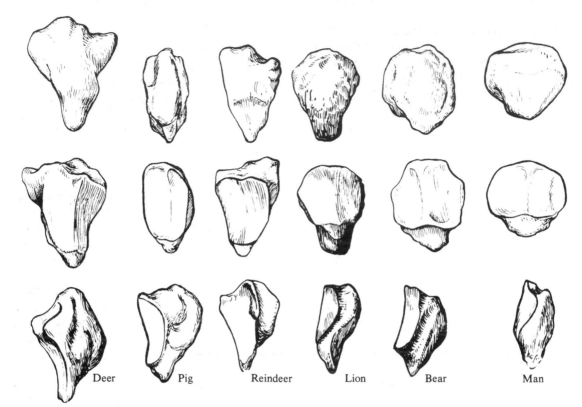

Figure 2.22   Mammalian patellae.

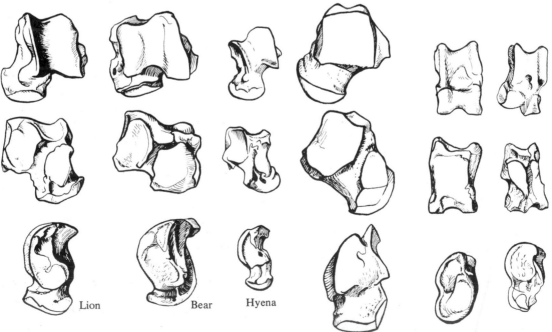

Figure 2.23   Mammalian tali.

differences (seen clearly in the distal views).

In the case of the radius, it is possible mistakenly to assign fragments to man, when they are in fact another species – especially a larger carnivore. The radius head may look superficially similar, but is in fact extremely round in man. The ulna is not so easy to confuse, although small pieces of shaft could be. The proximal articular end in man does not have an elongated olecranon process, and the articular surfaces are somewhat distinctive.

In the case of the tibia, parts of the shaft and articular ends could be confusing. In Fig. 2.21, superficial similarities at the proximal end can be seen, but there are generally more differences at the distal end, although it can be seen that the lion has slightly more of a human shape in that region. It is important to be careful in assigning species identifications to fragments of the proximal articular surfaces. The anterior border of the shaft is well defined along most of its length in man, and the bone is straighter. Shape of the shaft in cross-section can at times be useful in identification.

f) The patella

Some bones, such as the patella, can be seen to vary noticeably in size and shape in the larger mammals, and it will be seen in Fig. 2.22 that there is little overlap in appearance. There is, however, still a need to be careful in the case of fragmented bone, especially if there is added surface erosion.

g) The bones of the hand and foot

There can be some confusion in identifying correctly tarsal and metatarsal bones. Considerable shape differences occur in these elements, as seen by reference to the talus or astragalus (Fig. 2.23), where the carnivores are perhaps most like man.

Phalanges can perhaps give the most trouble, especially if eroded and incomplete. Fig. 2.24 gives some idea of the variation seen, by reference to a few species. The confusion is more likely to result from the similarities found in man and some carnivores. Included in the figure is a phalanx from Pin Hole Cave that was first identified and published as human, but which is in fact from a fossil lynx.

3   Cranial sutures

At the point where the growing edges of two skull bones come together, a thin membrane between them may persist unossified for some years after adulthood has been reached, or even indefinitely. This form of union is called a *suture*, and the bone margins may be flat and abutting, saw-edged or bevelled and overlapping. This joint must be differentiated from a *synchondrosis* which is formed when a larger residual plate of cartilage persists unossified between two bones. In man, the basi-sphenoid synchondrosis usually closes by about the twentieth year.

i) Sutures and ageing

Except in grossly abnormal individuals, the sutures usually begin to close at about 20

years of age, and may become obliterated later. The internal and external surfaces may not show similar degrees of fusion. As early as 1890, Dwight showed the possible use of suture closure in estimating age, and later Parsons & Boc (1905) put forward similar evidence. Todd & Lyon (1924, 1925) greatly elaborated on earlier work, considering that even sections of a suture could yield reliable age estimates. More recently, this work has come under criticism, particularly by Singer (1953b), Cobb (1955), McKern & Stewart (1957) and Genoves & Messmacher (1959), who consider that even if there is a general trend in suture closure, it is of little use as a guide for age determination. However, it is evident that in skulls with the facial region missing, closed or partially closed sutures at

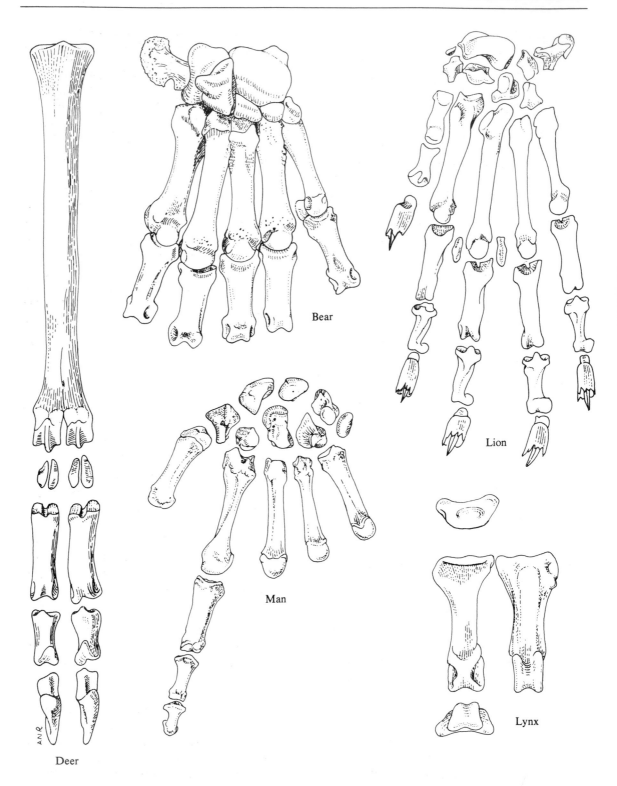

Figure 2.24   Mammalian phalanges, with the Pin Hole Cave phalanx of a lynx at lower right.

least show that the individual was adult, and this information is important when undertaking a metrical analysis. On the other hand, if the sutures are open, this need not necessarily indicate that the individual was under 20 years old (although the basi-sphenoid synchondrosis is a valuable indicator here). Sutures may also be of help in determining the number of individuals represented by a series of skull fragments. For example, if we compare a fragment showing a part of the sagittal suture nearly completely obliterated, with others displaying the suture completely open, it may usually be concluded that more than one person is represented.

ii) Abnormal sutures

a) Congenital anomalies

Under this heading fall the extra sutures that may be found at birth. In most cases these probably result from the independent existence of ossification centres that usually become fused before birth. As yet, no family studies have been attempted, and thus it is not known whether we are dealing with the effect of a single gene.

Perhaps one of the most well known anomalies is the *os japonicum* (Fig. 2.25). This is produced by the subdivision of the malar bone by a suture passing outwards from the lower margin of the orbit. It is usually seen in both malars, although the cases are still too few to be certain of this. It is an uncommon condition, with a frequency probably not greater than two or three in a thousand skulls. Thus, in large collections of Egyptians skulls Elliot Smith & Wood-Jones (1910) found only seven cases. Occasionally, on the palatal surface, a pair of sutures may be found running parallel to the median palatine suture and either anterior or posterior to the transverse palatine suture. The resulting extra bones are called anterior and posterior medio-palatine bones (Woo, 1948). The frequency of this anomaly may be no greater than that of *os japonicum*.

Much more has been written on the occasionally bipartite condition of the parietal. This 'twin' condition clearly results from a disorder of the parietal growth centres, and instead of the two primary centres of ossification joining during prenatal development to produce one common growth area, the centres remain separated by a suture (Fig. 2.26). As early as 1897, Dorsey noted about twenty cases in the literature, and this was followed in 1903 by three other compilations (Hrdlička, Le Double, Schwalbe). Since then more cases have been found, some being in excavated material. Kaufmann (1945), for example, notes the division of the left parietal in an ancient skull from Gland, Switzerland, while Goodman & Morant (1940) found traces of a divided parietal in one of the skulls from Maiden Castle, Dorset. As in a Papagattos Indian skull (Rusconi, 1940), the bipartite division may be associated with wormian bones.

b) Post-natal disorders

Excluding the major sutural anomalies of the skull, which will be described separately, a number of minor ones should be borne in mind. These may be caused by:

1) The retention of a feature found in early childhood. The metopic suture is particularly common, and has received full description in the section on non-metrical characters (p.90). More rarely, the exoccipital segments may remain clearly demarcated, at least from the upper (supraoccipital) part of the occipital, until adulthood.

2) Premature obliteration. Sometimes, even in young children, part of a suture may be obliterated long before the due time, and if skull growth is far from complete, deformity may result. Exactly what causes this early fusion is not known, although injury and disease may be contributory factors. Not only may the coronal, sagittal and lambdoid sutures of the vault show this irregularity, but sometimes the squamo-parietal.

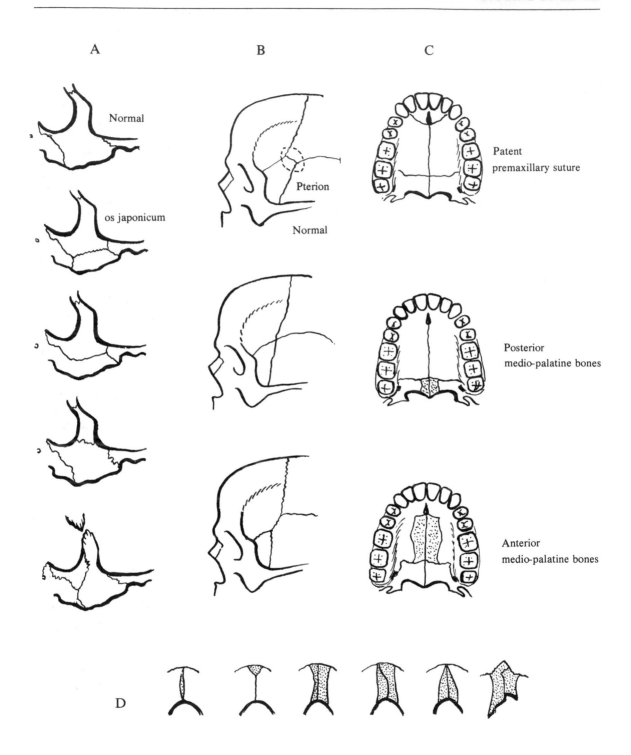

Figure 2.25   A, Varieties of the sutures and articulations to be found at the zygomatic arch. B, The most commonly found types of articulation at pterion. C, Bone and suture variations of the palate. D, Variations in the size and shape of the nasal bones.

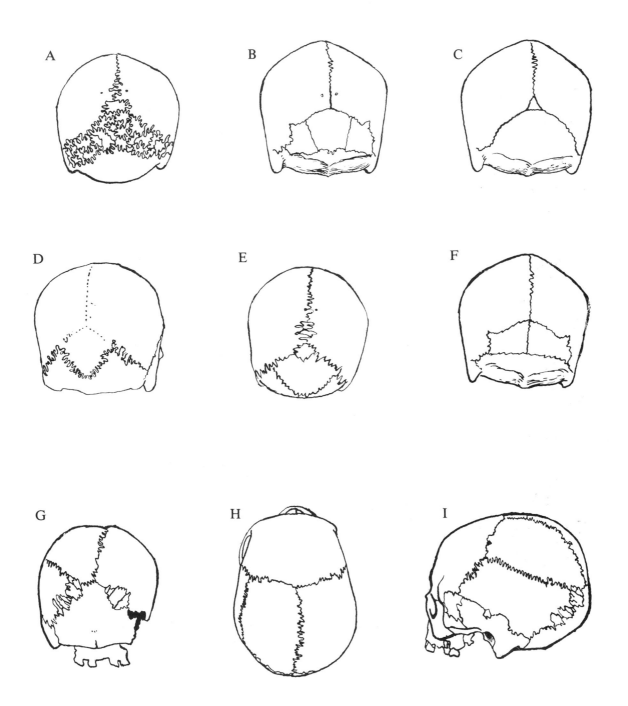

Figure 2.26   A–F, Bones and sutural variations to be found at the lamboid suture. B, tripartite Inca bone. F, Bipartite Inca Bone. G–I, Ancient skull from Switzerland displaying a bipartite left parietal.

4   Deformation of bones

The distortion of bone from the shapes usually encountered in a population may be due to one of three major factors.

i) Pathological causes

Although this section will be dealt with more fully later, it may be mentioned that certain anomalies with a genetic basis, such as *osteogenesis imperfecta*, or others caused by dietary deficiency, such as rickets, can produce gross bone deformities.

ii) Earth pressure

The degree and probability of deformation due to earth pressure may vary considerably with the soil and climate. In the case of the cranium, fusion of sutures helps to consolidate the brain-case and thus prevents distortion to some extent. Once the flat bones of the skull are broken or separated in the soil, the likelihood of distortion is far greater, especially of the parietals. The mandible, particularly if edentulous, may also be subject to lateral bending, which may become obvious when applying the condyles to the glenoid fossae. Apart from the fibula most post-cranial bones are not usually deformed by soil pressure to any noticeable extent, although in fossil material it is more frequently found (the distorted pelvis of *Australopithecus* from Sterkfontein is an example). It may be noted here that the skull is normally asymmetrical (Woo, 1931) and the more severe degrees are called plagiocephalic. Care should be taken not to mistake this for a post-mortem change of form, though both conditions may occur in the same specimen.

iii) Artificial deformation

a) Unintentional

Few bone deformities are known to result from accidental application of pressure to bones. In the case of post-cranial remains, one of the few examples is relatively modern: that is of rib-cage constriction through purposeful reduction of the diameter of the waist in women during the past few centuries (Flower, 1898). In the case of the skull, some instances of mild distortion may have resulted from over-tight bandaging. In the Americas, the widespread use of a cradle board has resulted in various degrees of unintentional cranial deformation.

There is some debate as to whether deformation may have occurred even in late Pleistocene groups, and particularly in some early Chinese, African and Australian specimens (Brothwell, 1975). This is, however, disputed (Larnach, 1974a) and clearly deserves further evaluation.

b) Intentional

Not many cases of post-cranial deformity of an intended character are known. The classic example is the distortion of the feet among Chinese women (Fig. 2.27), but this is a comparatively recent practice. Skull deformation is much more widespread, and there is a considerable literature on the subject which has been reviewed by Dingwall (1931). More recent studies are by Blackwood & Danby (1955) and Nemeskéri (1976). The changes in form are restricted to the braincase and are mainly the result of some type of antero-posterior compression. Artificial skull deformation has been most common in the Americas. Apart from the widespread deformation due to strapping the head to a cradle board in infancy, a number of tribes in ancient and modern times have practised binding the head to alter its shape in various ways (some Pre-Inca skulls from Peru and skulls of the Coast Salish and Kwakiutl tribes on the northwest coast are particularly noteworthy). Cases of artificial deformation of the head are also known in Europe, some dating from as far back as the Neolithic period. A few examples have been reported amongst British archaeological material.

The value of deformed skulls

Although deformation may restrict the number of metrical recordings that can usefully be made on a bone, and especially length, breadth and height of the skull (Helmuth, 1970), it need not render the specimen completely useless. The investigator will have to decide which observations may still be made according to the degree of distortion. In the case of the skull, S_1, S_2, S_3, FL, FB, G'H, G'$_1$, G$_2$, SC, DC, DA, H$_1$, M$_2$H and RB[1] are the dimensions least likely to be affected (see 'Points and "landmarks" of the skull', p.79).

Even when the skull is completely broken and distorted, it can still yield important non-metrical data, such as the number of wormian bones (see p.93), metopism, the number of caries cavities, the degree of dental wear, signs of injury and osteoporosis. Some post-cranial remains are always valuable for ageing and sexing.

In an extensive analysis of Indian skull material from Pecos Pueblo, Hooton (1930) employed a statistical correction for cranial deformity. However, he points out that average values only may be obtained, and such formulae are useless for the prediction of individual values.

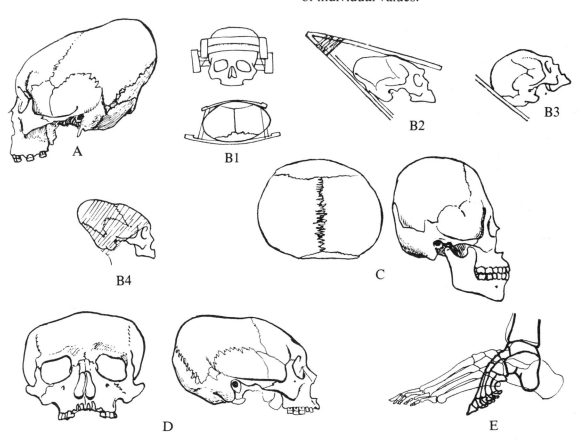

Figure 2.27   Cultural deformation. A, Kwakiutl skull from British Columbia with vault deformity through binding. B, Methods of producing vault deformity: 1, antero-posterior compression. 2, special cradling device (Chinook). 3, occipital flattening. 4, binding. C, Top and lateral views of a Mexican skull showing antero-posterior flattening. D, Lateral and facial views of a skull showing deformity due to vertical compression. E, Skeleton of a deformed Chinese foot superimposed on normal foot bones.

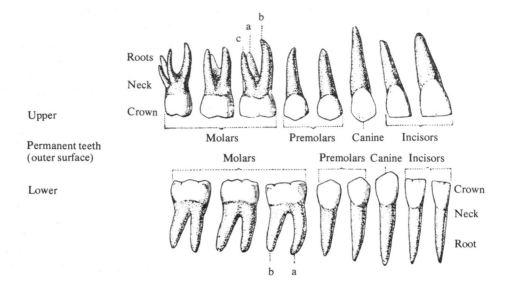

Upper

Permanent teeth
(outer surface)

Lower

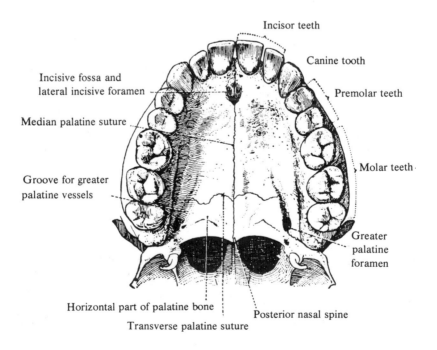

Figure 2.28   Human teeth and the adult palate.

5 · The teeth

All human teeth display three well-defined portions (Fig. 2.28):

a) the crown, which is the part of the tooth situated above the gum;

b) the neck, a slightly constricted portion immediately below the crown;

c) the root, which is the rest of the tooth below the neck, and which is enclosed in the tooth socket.

It need hardly be mentioned that the teeth are arranged in a semi-circular arch, with part of each tooth directed towards the lip or cheek (known as the labial or buccal surface), and another part directed towards the tongue (lingual surface). In addition, the incisors, canines, premolars and molars have mesial and distal surfaces. The upper and lower teeth do not always fit together (occlude) in the same way in all individuals, with the result that various types of occlusion have been defined (Dickson, 1970). Badly fitting upper and lower teeth (noticeable malocclusion) are not common in human remains from Europe of more than five centuries old, or for that matter in modern primitive groups such as the Australian aborigines. However, the slight rotation and deflection (Fig. 2.29) of one or a pair of teeth is not so uncommon in late prehistoric material, and this can produce a slightly anomalous occlusion. Marked overbite or underbite (Fig. 2.30) are extremely rare in the earlier remains.

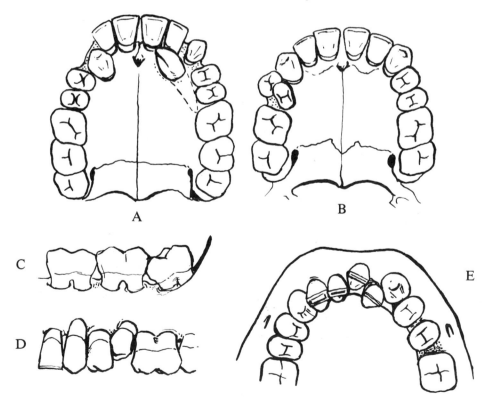

Figure 2.29   Anomalous positions of teeth. A, Malpositioned upper canines, with the retention of the milk canine on the right. B, Irregularly placed and rotated premolars on the left of the palate. C, Impacted third molar. D, Impacted second premolar. E, Crowding and rotation of the anterior teeth to the right of the jaw.

Man develops two sets of teeth, although the frequency of dental disease in the civilized world would make a third very welcome. No teeth are visible at birth, but by the end of the first year a number have erupted. These deciduous or milk teeth (Fig. 2.31) are later replaced by the permanent set. Generally, the teeth in both sets erupt in a definite order, as shown in the chart on dental eruption (p.65). Times of eruptions also tend to be fairly constant, although there can be considerable variability (Garn *et al.*, 1959).

In both the deciduous and permanent dentitions, there are a number of groups of four teeth of similar form and occupying similar positions in both sides of the upper and lower jaws. For example, there are four permanent canines, two in the upper jaw (left and right) and two in the lower (left and right). The fact that they are the third tooth from the front of the mouth is shown in the dental formula for permanent teeth, where they are given the number three in each quadrant. The formula is therefore written thus:

Rt. 8 7 6 5 4 3 2 1	1 2 3 4 5 6 7 8 Lt.	Palate
8 7 6 5 4 3 2 1	1 2 3 4 5 6 7 8	Mandible

The numbers refer to the following teeth:

1 = medial incisor   5 = posterior premolar
2 = lateral incisor   6 = first or most anterior molar
3 = canine   7 = second molar
4 = anterior premolar   8 = third molar or wisdom tooth.

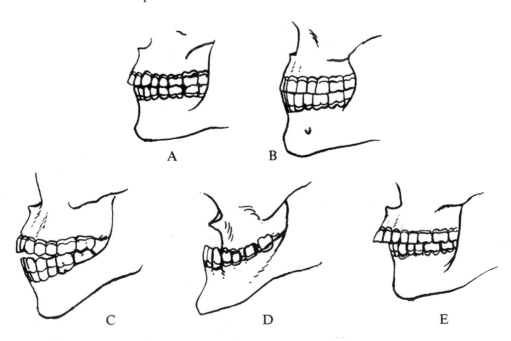

Figure 2.30 Various types of dental occlusion to be seen in man. A, Most common with slight overlap of the upper anterior teeth over the lower ones. B, Edge-to-edge bite of the anterior teeth. C, Anterior open bite (rare). D, Extreme protrusion of the lower jaw and teeth. E, Marked overjet of the upper anterior teeth.

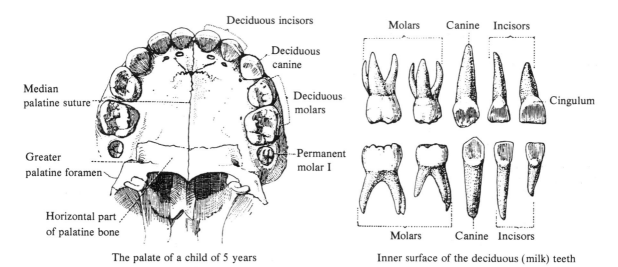

Deciduous incisors

Deciduous canine

Median palatine suture

Deciduous molars

Greater palatine foramen

Permanent molar I

Horizontal part of palatine bone

The palate of a child of 5 years

Molars   Canine   Incisors

Cingulum

Molars   Canine   Incisors

Inner surface of the deciduous (milk) teeth

Figure 2.31   The palate of a 5 year old child and the inner surface of deciduous (milk) teeth.

This formula may at first seem rather complicated, but in fact the notation is easy to master and allows one to record what teeth are present with far greater rapidity. When recording, the jaws should always be in the same position, and it is the practice of the author to examine the palate with the teeth forming an arch, and the mandible with the teeth forming a loop (in other words, with the left side of the jaws at the right hand).

Now that a general formula for the permanent dentition has been devised, other symbols can be used to denote further information. The following are suggested as basic symbols although it does not matter what signs are used provided that they are defined in the report (printers appear to have varying preferences).

area missing

8 7 6 5 4 3̶ 2 1 | 1 2 3 – – – – –
\8̷ 7 8̸ 5 4 3 2 1 | 1̶ 2 3 4 5 6 7 ⑧
                          C A

X̶ = tooth missing but socket present

3̶ = tooth present but socket missing

8̸ = tooth lost ante-mortem

\8̷ = tooth not yet erupted

⑧ = tooth probably erupting

5 = tooth has a caries cavity
C

6 = tooth displays an abscess at the root.
A

In order to differentiate clearly between deciduous and permanent dentitions, letters of the alphabet are used for the former. Thus we get:

Right e d c b a | a b c d e   Left part of palate

Right e d c b a | a b c d e   Left part of mandible

a = medial milk incisor
b = lateral milk incisor
c = milk canine
d = first milk molar
e = second milk molar

Modifications of these formulae may be used when one set of teeth is being shed, or when a milk tooth is retained with the permanent dentition. For example, we might get the following:

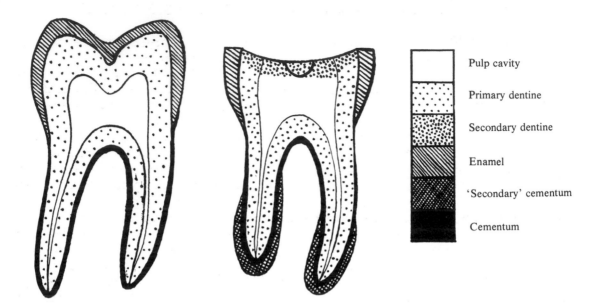

Figure 2.32   Diagrammatic sections through a worn and unworn human molar, showing the various dental tissues.

a) Dentition of a child with the permanent teeth beginning to erupt:

$$\overline{\underline{\backslash 7\, 6\ e\ d\ c\ 2\ 1\ |\ 1\ 2\ c\ d\ e\ 6\ 7/}}$$
$$\backslash 7\, 6\ e\ d\ c\ 2\ 1\ |\ 1\ 2\ c\ d\ e\ 6\ 7/$$

b) Dentition of an adult with one milk tooth still in position:

$$\underline{8\ 7\ 6\ 5\ 4\ 3\ 2\ 1\ |\ 1\ 2\ 3\ 4\ 5\ 6\ 7\ 8}$$
$$8\ 7\ 6\ 5\ 4\ c\ 2\ 1\ |\ 1\ 2\ 3\ 4\ 5\ 6\ 7\ 8$$

3   milk canine still in position

permanent canine retained within the jaw.

If there is any possiblity of confusion the symbol for a milk canine with that denoting a caries cavity, then it is best to clarify this point in writing.

All human teeth display the same layers of dental tissue (Fig. 2.32). The crown is covered with an extremely hard substance called enamel, which may occasionally extend on to the root (for example, as an 'enamel pearl', Fig. 4.21). The tissue beneath the enamel, forming also most of the root, is called dentine, and is relatively much softer. Within the dentine is the pulp cavity extending from the crown into the roots. Covering the dentine at the root is the cementum, which is an even softer tissue. During adult life, additional cementum may be deposited on the external surface at the base of the roots, either as a result of oral disease or as a part of the ageing process. Finally, in teeth that become extremely worn, it is usual to find 'secondary dentine' occupying the area of the exposed pulp cavity. As yet, there are remarkably few detailed general studies of teeth in archaeological series, two of the exceptions being the Coxyde population (Twiesselmann & Brabant, 1967) and medieval Danes (Lunt, 1969).

6   Value of X-rays in studying bones and teeth

X-rays are specialized waves that are able to penetrate many substances. Their penetration of the different tissues of the human body varies in degree. Bone, containing calcium salts, is resistant and therefore can give a clear picture not only of the external form but also of the internal structure. An X-ray photograph (known as a radiograph) of the side view of a skull, for example, gives a fairly clear sagittal view as well as revealing other bone contours; similarly when a femur is portrayed on X-ray film, the internal structure of the bone is revealed, including the lines related to the stress directions (Fig. 1.2)

Although instruction from a qualified radiographer will be needed in this field of study, the value of radiography should not be forgotten in preparing preliminary reports on bones. Certain conditions may be suspected from a surface examination but need confirmation. Sometimes it is difficult to be sure whether a bone swelling is the site of a healed fracture, and X-raying can help to confirm the diagnosis. A swelling may also be, for example, the result of 'Albright's disease', periostitis or a tuberculous abscess within the bone. Radiography will also show whether the swellings in an osteitis or an osteoma are associated purely with the outer layer of bone, or whether they have also affected the marrow cavity.

In the case of skull thickening, X-ray examination should help to differentiate between such pathological conditions as Paget's disease (Pl. 5.2), various anaemias and thickening associated with other factors.

If it is suspected after external observation that the frontal, maxillary and sphenoid sinuses may be smaller than normal, radiography should give confirmation or the reverse.

Dental radiography is important in estimating the exact age of non-adult skulls. It also facilitates the detection of any unerupted teeth. Suspected caries cavities, especially where there is only a small pit to be seen in the enamel, may be seen through radiographs to extend into the dentine.

The common positions used in X-raying the skull are clearly shown in Fig. 2.33, but positioning may need to be modified in relation to the nature of the investigation. Parts of long-bones, pelvis or other post-cranial bone can be filmed according to what information the archaeologist wishes to gain. The great range of information to be derived from the radiographic study of archaeological materials is reviewed in Wells (1963), Brothwell, Molleson and Metreweli (1968) and Brothwell, Molleson, Gray and Harcourt (1969). Examples of the type of work that can be undertaken using X-rays are seen in the studies of the frontal sinuses in

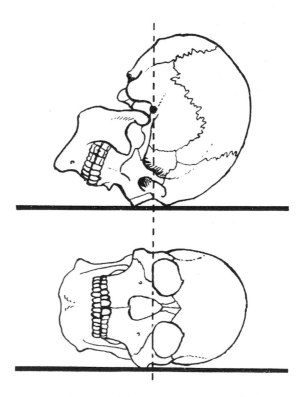

Figure 2.33   Standard positions for obtaining facial and lateral radiographs of the skull. The thick horizontal line represents the position of the X-ray film.

neanderthalers (Vlček, 1967), early British groups (Buckland-Wright, 1970), and maxillary prognathism in Greenland eskimo skulls (Krogstad, 1972).

The archaeologist should find no great difficulty in getting specimens X-rayed. Such work is most easily accomplished through a museum or medical school, although as regards dental-age estimations local dentists are often helpful.

It may be noted that X-ray duplicates can be obtained with a minimum of inconvenience (Armour-Clark, 1957), so that specialists wishing to have records of bone anomalies can usually do so without keeping the original plates.

7 Microscopy and electron microscopy

Although the microscopy of mummy tissue has a long history back to the studies of Sir Marc Armand Ruffer early this century, the study of bone under higher magnification has been somewhat neglected. In view of the need for some magnification in the study of osteones in relation to age (Kerley, 1965; Ortner, 1975; Ubelaker, 1974) this is clearly becoming an established technique. There may also be value, when considering certain kinds of pathology and in preparing and studying sections of bone (Blumberg & Kerley, 1966; Ascenzi, 1969). Morse (1969) for instance supports a diagnosis of Amerindian bone disease with section details. Electron microscopy, again, has been applied to mummy tissue (Macadam & Sandison, 1969) but clearly has wider application.

When wishing to view (and record photographically) higher magnifications, the scanning electron miscoscope is especially of value (Pl. 2.1). Its versatility is such that excellent magnification of bone surfaces can be from × 100 to over × 10 000. As yet its potential has still to be evaluated, but it may well assist in certain problems of the differential diagnosis of bone disease. It could also assist in studies on the deterioration and fossilization of bone (Imai, 1970; Hermann, 1972; Day & Molleson, 1973). Microradiography is a further technique that may well prove of value in considering the biology of ancient bone (Sergi, Ascenzi & Bonucci, 1972; Stout, 1978).

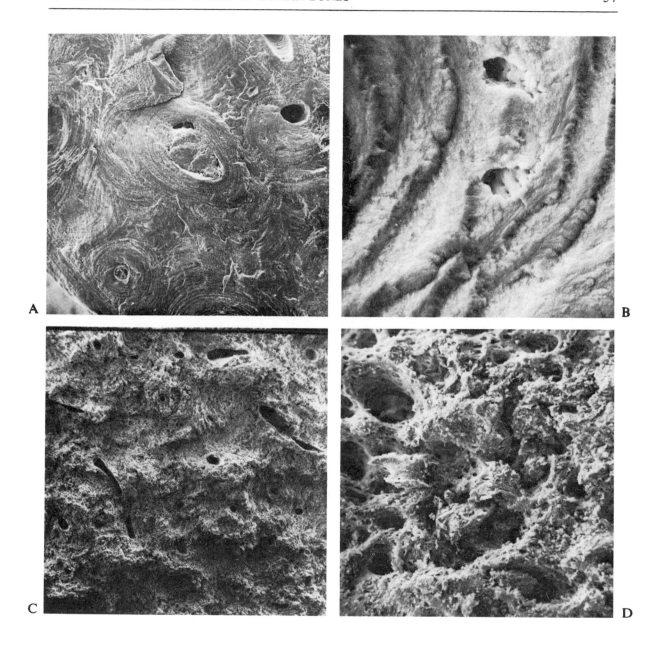

Plate 2.1   Scanning electron micrographs of bone. A, TS of fossilised long-bone from Kedung Brubus showing Haversian systems and inner circumferential bone ×120. B, TS of fossilised long-bone from Trinil Femur III showing bone lamellae and osteocyte lacunae in a Haversian system ×1250. C, TS of unfossilised long-bone from Winchester medieval cemetery showing Haversian systems ×75. D, TS of a modern cow bone with industrial fluorosis showing a Haversian system and disorganised and porous bone mineral structure ×150.

III Demographic aspects of skeletal biology

1 Sexing skeletons

Before discussing specific features used in determining the sex of an individual, a number of general points must be emphasized. First it must be appreciated that the value of certain features varies with the group being studied. Thus, the degree of supraorbital development that serves in Europeans to identify males, may be found in a number of females among Australian aborigines. Similarly, the robustness of female Australian aboriginal bones may be far greater than is generally found in male pygmies.

Secondly, it must be remembered that both as regards measurements and general shape, there is often considerable overlap in the range that is found in the two sexes. Yet another complicating factor is the frequent incompleteness of skeletal remains with the result that sex may have to be determined provisionally on only one of two features.

Ideally, it is necessary to have a large series of fairly complete skeletons of one particular ethnic group, for it is in most cases only by noting the variability in the collection that any degree of certainty in sexing can be achieved. With much archaeological material, there is a constant danger of incorrect sexing, and indeed, Weiss (1972) suggests that there is a 12 per cent bias in favour of males. For convenience, the various parts of the skeleton will be considered separately.

As a result of a European meeting in 1972, recommendations on sexing skeletons have been published in the *Journal of Human Evolution* (1980) **9**, 517-549. These are worth considering in conjunction with the comments made below.

It should be noted that in the case of very fragmentary bones, it may still be possible to sex the individuals by the chemical evaluation of citrate in the bones (more being generally present in the female). So far, the studies by Kiszely (1974), Dennison (1979) and others suggest great promise for this technique.

The skull

This is not the easiest region of the skeleton to sex, especially if broken and fragmentary. Keen (1950) has discussed sex differences of the skull in some detail, while Garn, Lewis, Swindler & Kerewski (1967) and Garn, Lewis & Kerewsky (1967) consider sexual dimorphism in tooth dimensions. Also, Ditch & Rose (1972) show that the multivariate statistical analysis of dental measurements may aid in the sexing of skeletons by other features. In general the male skull may be distinguished from the female by the following characteristics (Fig. 3.1):

a) It is generally larger and heavier.

b) Muscular ridges, such as the temporal lines and nuchal crests, are larger.

c) The supraorbital ridges are more prominent and the frontal sinuses larger.

d) The external occipital protuberance and mastoid processes are more developed.

e) The upper margin of the orbit is more rounded.

f) The palate is larger.

g) The teeth are often larger (mesiodistal and buccolingual crown diameters).

h) The posterior root of the zygomatic process extends for some distance past the external auditory meatus as a well-defined ridge.

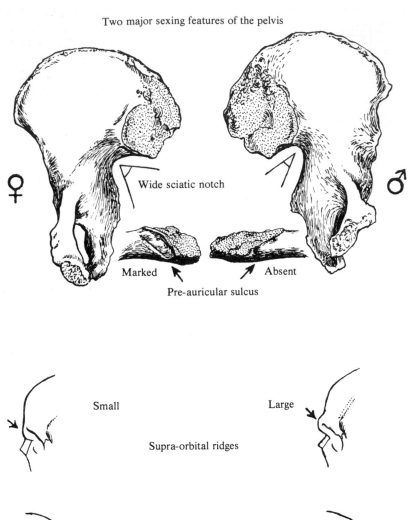

Two major sexing features of the pelvis

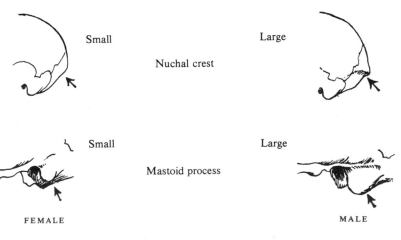

Figure 3.1   Principal sexing features of the pelvis and skull.

i) The mandible is more robust with more developed and flaring gonial regions.

j) The ramus of the mandible is broader and longer in males with a better-developed coronoid process.

k) Lastly, it may be noted that the male skull is less rounded, the female skull tending to retain more the adolescent form.

In contrast to these visual methods of assessing sex, it should be stressed that where large numbers of skulls are involved, and computer assistance is available, sex determination by discriminant function analysis has proved to be very rewarding (Giles & Elliot, 1963; Giles, 1964; Birkby, 1966).

Vertebral column

a) In males the general size, especially in the lumbar region, is greater.

b) Overall ruggedness and the extent of markings for muscle and ligament attachments are more in evidence.

c) The estimated total length of the spinal column is greater. Cunningham (1951) and Smith & Fiddes (1955) find a mean difference of about 100 mm between the sexes, for total length. A rough estimate of column height can be made by adding together the vertebral-body heights (excluding sacrum and coccyx). Such estimates, however, cannot be considered entirely reliable (Harrison, 1953).

d) The atlas vertebra is distinctly more massive in the male (Boyd & Trevor, 1953) with a number of dimensions, particularly the breadth, greater than in the female.

e) The sacrum is longer and narrower in the male, this difference being clearly expressed by the sacral index. Various measurements can in fact be used to demonstrate differences between males and females (Flander, 1978).

f) The abdominal aspect of the sacrum in males displays a uniform curve with the deepest part of the hollow at the third segment; whereas in females, the upper portion is flattened and the lower portion sharply angulated (the deepest part being at the fourth segment).

g) The auricular surface (articulating with the ilium) is limited to the first and second sacral vertebrae in the female, but in the male it often extends to the middle of the third vertebra.

Sternum

In the male, the body is at least twice as long as the manubrium, while in the female, it is relatively shorter. This fact probably has over a 50 per cent accuracy. In European sterna, the combined midline length of manubrium and mesosternum usually equals or exceeds 149 mm in males, but is usually less in females (Ashley, 1956).

Clavicle

Although this bone does not yield any absolute sexing criteria, it may be noted that the male clavicle is generally more robust and on average over 10 mm longer than the female (Parsons, 1916).

Scapula

The scapula of the male differs on an average from that of the female, in size, proportions and shape. Probably the most useful study so far is by Bainbridge & Genoves (1956). Oliver & Pineau (1957) have also undertaken a metrical analysis of male and female scapulae (presumably using European material). This biometric method is of value in specialist studies but is probably of less value in general reports, especially as the scapula rarely survives intact.

Pelvis

There is no doubt that this bone yields the most reliable sexing information of all, and it is probable that between 90 and 95 per cent accuracy in determination can be achieved (Krogman, 1946; Washburn, 1948; Genoves, 1959; Phenice, 1967).

Many features of the pelvis have been stated to show sex differences, but only those that are particularly important and can be described easily are mentioned here.

The characters used may be divided into two groups: (i) those depending upon visual examination, (ii) measurable dimensions.

i) By far the most important diagnostic features for the excavator are those that can be sexed upon inspection. These differences are related to the fact that the female pelvis is specially adapted for childbirth, with the result that there is more accommodation within it than in the male, while the relative depth is less. The morphological points to note are as follows:

a)–d) of minor value

a) As a whole, the male pelvis is more robust, with well marked muscular impressions.

b) The depth of the pubic symphysis is generally greater in the male.

c) The acetabulum is larger in the male.

d) The obturator foramen is larger in the male and rather oval in outline, whereas in the female it is smaller and more triangular in shape.

e) and f) of great value

e) The sciatic notch is narrower and deeper in the male (Fig. 3.1). Although in all groups this is an excellent diagnostic feature, variations in the general shape of the notch do occur. Genoves (1959), for example, found that its general shape in Anglo-Saxons was quite different from that in British medieval individuals of the same sex.

Even in non-adults the notch has been claimed to have some utility. Imrie & Wyburn (1958) consider the notch difference to be 'inborn', and Thompson (1899) noted that it is larger even in the female foetus. This dimorphism has also been claimed in the infant (Reynolds, 1945, 1947), but this is doubtful. Recently, Weaver (1980) has suggested that foetal and infant ilia show a 91 per cent accurate characteristic in the so-called 'auricular surface deviation.'

f) The pre-auricular sulcus (Fig. 3.1) is more constantly present in the female ilium, although sometimes poorly developed or present on one side only. This groove has been called an 'inborn' difference (Imrie & Wyburn, 1958). On the other hand, Houghton (1974) considers that there are two forms of 'groove', one marking the attachment of the inferior part of the ventral sacro-iliac ligament, the other being purely feminine and the result of changes that can occur at the site of the attachment of the pelvic joint ligaments during pregnancy and childbirth.

ii) Differences between some parts of the male and female pelvis have also been described in metrical terms. The important ones may be stated briefly as follows:

a) A sub-pubic angle of 90° or over usually indicates the female sex.

b) The ischio-pubic index (calculated from ischial and pubic lengths as defined by Schultz, 1930) is lower in the male. This index was devised to replace the sub-pubic angle.

c) The angle of the sciatic notch (estimated on a shadow tracing) is much smaller in the male. This means of discrimination takes the place of visual examination of the area.

The value of the last two characters is seen clearly in the scatter diagram produced by Hanna & Washburn (1953) for Eskimo pelves (Fig. 3.2). However, the landmarks for these measurements are somewhat ill-defined and may lead to inaccuracy (Stewart, 1954; Thieme & Schull, 1957). With the definition of more precise points, metrical procedure for sexing will be valuable in the study of large series, especially as measurement lends itself to more reliable statistical analysis (as Pons, 1955*a,b*; Thieme, 1957; Thieme & Schull, 1957 and Howells, 1964, have shown). To what extent such metrical methods will be used only in specialized analyses remains to be seen, for as Stewart (1954) suggests, it may be rather a waste of effort to measure such specimens simply to verify what can so quickly be seen by eye. For general purposes the 'visual' features can be relied upon for sexing the pelvis.

In the case of detailed studies on fossil hominid pelvic fragments detailed measurements and the application of multivariate statistics might be useful in sexual differentiation (Day & Pitcher-Wilmott, 1975).

Long-bones

Hrdlička sums up the value of the long-bones in sexing by saying that in the male they are longer, heavier and have larger attachment areas for muscles (including the linea aspera, crests, tuberosites and impressions) (Stewart, 1947). Moreover, he considered that the most important and constant sexual difference lies in the regions of articulation. These comments certainly apply if a series is being sexed. But if only one or two long-bones (or fragments) are available, it may be difficult to draw any conclusions unless the bone displays very clear indications of maleness or femaleness (Boyd & Trevor, 1953).

Vallois (1957) considers that the weight of a long-bone (which takes into account the size-differences of various areas) to be a better discriminant than other long-bone characters which have been employed. However, the value of weight in archaeological material seems debatable, for one would have to take into account

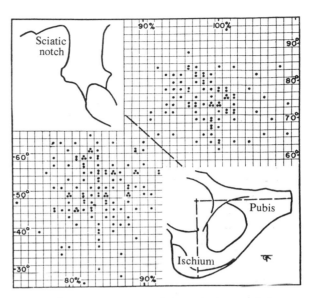

Figure 3.2   The angle of the sciatic notch (see Fig. 3.1) in degrees plotted against the ischium-pubis index in per cent. Diagonal line separates 'males' and 'females'.

differences in preservation, as well as the possibility of soil occurring within the bone.

Femur. As in the pelvis, sex has been judged both on morphological and metrical grounds. In small series of femora, and as supporting evidence of sex, visual observations are sufficient. In males, the bone as a whole is longer and generally larger, but especially the head and distal condyles; while the shaft is broader and (in section) thicker, with a more prominent linea aspera.

Metrically, a number of features have been considered of value. Parsons (1914), in a study of English femora, found the vertical diameter of the femoral head and the bicondylar width of the distal end to be the most reliable sexing dimensions. Pearson & Bell (1919) undertook a far more extensive analysis, but it seems probable that Parsons' original measurements are still the most reliable known, especially those of the femoral head diameter (Thieme, 1957).

In the case of poorly preserved and fragmentary material, Black (1978) found that the midshaft femoral circumference was a useful discriminant.

2   Estimation of age

Assessments of age based on skeletal remains are most likely to be fairly accurate with immature or young adult individuals. Remains of older persons present more of a problem, and when dealing with earlier populations, it is difficult to be sure that significant age-changes took place at the same time, and that they showed the same group variability, as in modern populations. Work on skeletal ageing is still far from complete; most of the ossification periods and eruption times have been worked out on American and European samples, and may

not apply exactly to other parts of the world's population. The most recent general survey of ageing is that of Genoves (1969).

Climate and diet have a considerable effect upon maturation, as for example the work of Weiner & Thambipillai (1952) on West African children suggests, and therefore any estimates of life-expectancy in earlier populations must be accepted with reservation.

As certain parts of the skeleton are more valuable for age estimations than others, it is as well to deal with these separately. Ageing by sutures, which has now fallen into

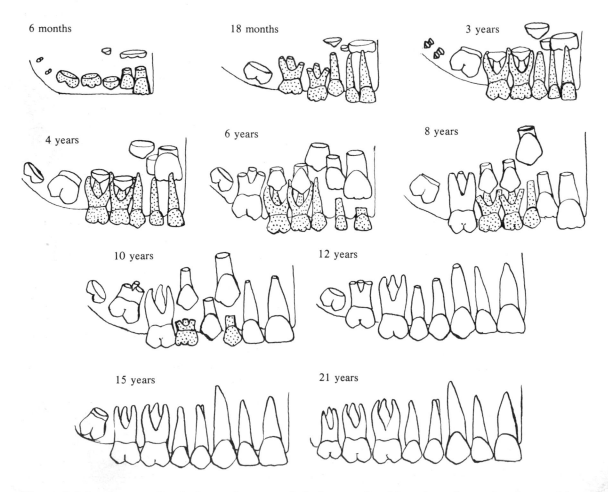

Figure 3.3A   Average developmental stages of the human dentitions from 6 months of age to 21 years. Stippled teeth represent the milk (deciduous) dentition.

disfavour, has already been discussed (p.43), although the recommendations of the 1972 European meeting (*Jl. hum. Evol.* 1980) do give a scheme for scoring age from degree of suture closure.

The skull

In abandoning the sutures as a method of ageing, few features of the skull except the dentition are of much use for our purpose. If the skull is noticeably thin and light, it may well be an immature specimen, though it is difficult from thickness alone to give an accurate age. In the case of very immature individuals, the ossification sequence of the occipital bone has been found to be useful (Redfield, 1970). Well-formed mastoid processes and external occipital protuberance, as well as frontal sinuses extending high into the supraorbital region, suggest

an adult individual. These features are not usually so well marked in females. In most individuals, the basi-occipital begins to fuse with the basi-sphenoid at about the 17th year, and they are usually completely joined by the 20th to 23rd year. There is also evidence to show that the sphenoidal sinus may extend into the occipital bone after the age of 25 (Dutra, 1944).

By far the most valuable age indicators in the skull are the teeth, not only during the course of eruption, but also as regards changes during adult life. The use of dental attrition or wear is discussed in another section (p.71). Rather than discuss here each tooth and its eruption separately, the times and general sequence of eruption are given in Figure 3.3. Although the times were mainly estimated from American data, they are nevertheless generally applicable. It should be noted that the average times are given in *A* but in fact there is a wide range of variation within most populations as seen in *B* of Fig. 3.3B (Garn & Rohmann, 1966). Evidence collected from various parts of the world suggests local differences in mean eruption times; even the deciduous (milk) dentition shows some divergences (Falkner, 1957).

Radiographs of immature jaws reveal clearly the state of development, although in broken remains, the unerupted teeth may be sufficiently exposed to eliminate the need for X-ray examination. Minor differences in average eruption times are to be found between the sexes, but unless special child studies are being made, this hardly seems worth considering, especially as both show considerable variability.

With regard to ageing adult teeth, Gustafson's (1950) work has helped to place such determinations on a relatively scientific basis. His method consists principally of assigning values to each of the following criteria: attrition, alterations in the gingival attachment, secondary dentine, thickness of the cementum, root resorption and translucency of the root. Unfortunately, not only would any attempt to use most of these criteria on

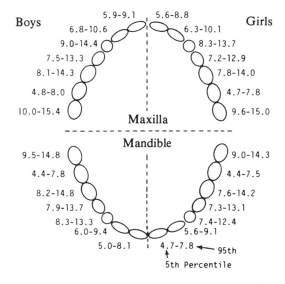

Figure 3.3B   Fifth and ninety-fifth percentiles for permanent tooth eruption in boys (left) and girls (right). Although tooth eruption is somewhat less variable than many aspects of skeletal development, the 'six-year molar' is properly the 4-, 5-, 6-, 7-, or 8-year molar, and the '12 year molar' is nearly the 9-, 10-, 11-, 12-, 13-, 14- or 15-year molar!

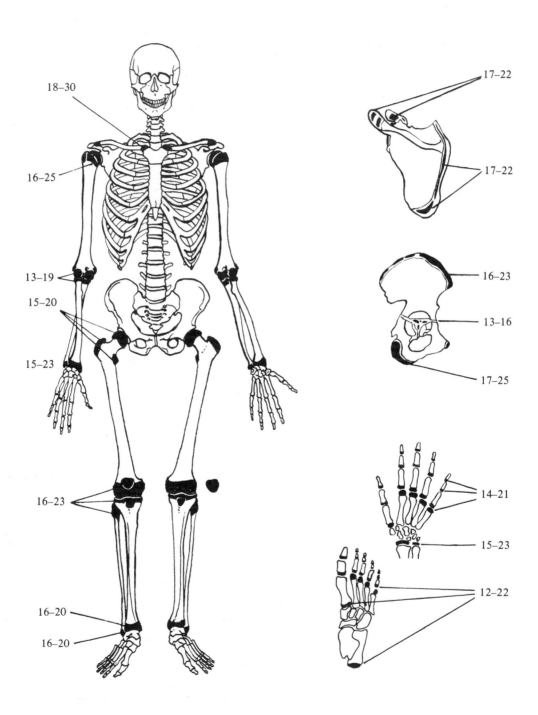

Figure 3.4 *The times of epiphyseal union of various parts of the skeleton. All numbers represent years, the difference between each pair showing the time span within which the particular epiphyses unite. Data from various sources.*

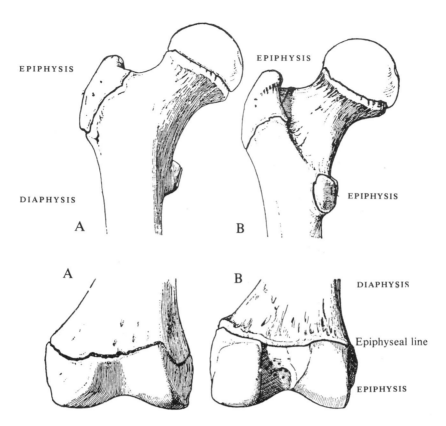

Figure 3.5   Anterior (A) and posterior (B) aspects, proximal and distal ends, of an immature femur. The epiphyses have not yet united and there is a well-marked epiphyseal line in each case.

early material be very laborious, but it would first be necessary to have teeth of known age from the same ethnic group and period to serve as standards for comparison.

The post-cranial skeleton

The ageing of post-cranial remains without supporting dental information is not always an easy matter, especially in preadolescent children. The majority of work on the bony development of such young individuals has been done on living series, which presents far fewer problems than when dealing with excavated skeletons. Remains of children are so often incomplete, owing to the extreme difficulty in excavation and the ease with which the small epiphyses and hand bones erode and disintegrate. In the case of the epi-

physeal area, it is often virtually impossible to compare excavated bones of children with the illustrations in the standard atlas by Greulich & Pyle (1950).

Therefore, in consulting the graphic summary of the normal progress of ossification (Fig. 3.4), it should be realized that the remains may fall very short of the quantity and quality desired. Particular attention should be given to finding and identifying the larger epiphyses (Fig. 3.5) as these are more likely to be preserved. The knee region is thus probably the most useful in ageing preadolescent skeletons (Hunt & Gleiser, 1955), and the best reference work in this connection is the atlas compiled by Pyle & Hoerr (1955).

Perhaps it is as well to remind the reader here of the nature of the epiphysis. An epiphysis may be defined as 'a cartilaginous area

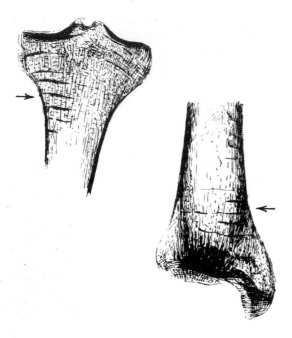

Figure 3.6 Radiographs of the proximal (above) and distal (below) ends of a pre-Roman tibia, showing well marked lines of arrested growth (arrowed).

present at each end of every long-bone of the limbs, on the upper and lower faces of the vertebral bodies and in certain other locations where special processes are required for the attachment of muscles' (Watson & Lowrey, 1951). Ossification begins apart from that of the main centre of ossification for the bone. Later, the epiphyses fuse with the main centre to form one solid bone. Sometimes it is possible to detect in radiographs transverse lines along the shaft of the diaphysis of long-bones, often called 'lines of arrested growth'. These may be related to illness in some cases; Harris (1933), for example, has described a typical example in pre-Roman human remains from a Derbyshire cave (Fig. 3.6). In recent years various population studies investigating these lines suggest that their occurrence and retention may be linked to diet, disease and growth, but that their analysis in environmental terms may not be easy (Marshall, 1968; Garn, Silverman, Hertzog & Rohmann, 1968).

The period of epiphyseal growth that concerns us in younger children, is replaced in the age period 12–25 years by epiphyseal union and final skeletal maturation. It cannot be too strongly emphasized that even within a group of normal individuals, considerable variation in maturation times are to be noted, and females tend to be slightly ahead of males. Another complicating factor is that mean differences are present between various groups throughout the world, determined by a combination of genetic and nutritional factors.

Variation also occurs in the development of the pubic symphysis, another area of bone much used in ageing. However, this does not lessen its value, and it is in fact more useful than the others in so far as the changes extend into the later decades of adult life. Assessing the age of a skeleton from changes of the symphyseal face was first studied in detail by Todd (1920, 1921) but he based too much on averages (Brooks, 1955). A very important contribution has since been made by McKern & Stewart (1957) who studied skeletal age-changes in young American males. They summarize the part of their work dealing with pubic changes as follows:

'Since Todd's system of typical phases serves only those symphyses that conform to his concept of typical, we have proposed a new method of ageing in which symphyseal metamorphosis is evaluated in terms of combinations of its component parts. Thus we have selected three components, each subdivided into five developmental stages which, when combined as a formula for any pubic symphysis, will yield an age range and the probable age of the individual. In comparison to Todd's system, the symphyseal formula expresses the true nature of symphyseal variability and does not confine the observer to the narrow limits of typical phases.'

Their method of obtaining an age may be described briefly as follows. Compare the symphyseal face with the stages in each of the three components as shown in Fig. 3.7, assigning to it three stage numbers. It should

Component I

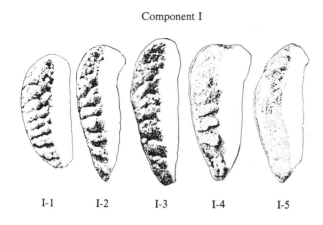

I-1   I-2   I-3   I-4   I-5

Component II

II-1   II-2   II-3   II-4   II-5

Component III

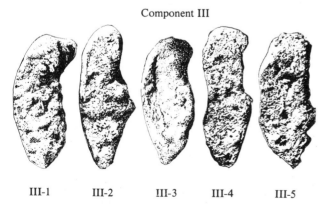

III-1   III-2   III-3   III-4   III-5

Figure 3.7   Age changes at the pubic symphysis.
The five stages of the three components are given
as described in the text.

be noted that in component I, only the dorsal half (demi-face) is considered, and the stage 0 represents the absence of the dorsal margin that delimits the demi-face. In component II, which comprises the ventral half, various degrees of bevelling and rampart formation are seen, and in stage 0 ventral bevelling is absent. Component III is characterized by the formation of a distinct and elevated rim surrounding the whole of the now level face, and in stage 0 this rim is absent.

Thus, one might find on comparing a specimen that stage 3 is only present in component I and 0 in the other two cases, thereby assigning the numbers 3–0–0. By referring to Table 1a, this formula can be translated into the modal frequencies of 20, 19 and 19 years (the mode being the most commonly occurring age for each age-range). From these

Table 1a

Age limits of the component stages

Stage	Age Range	Mode
Component I		
0	17.0–18.0	17.0
1	18.0–21.0	18.0
2	18.0–21.0	19.0
3	18.0–24.0	20.0
4	19.0–29.0	23.0
5	23.0 +	31.0
Component II		
0	17.0–22.0	19.0
1	19.0–23.0	20.0
2	19.0–24.0	22.0
3	21.0–28.0	23.0
4	22.0–33.0	26.0
5	24.0 +	32.0
Component III		
0	17.0–24.0	19.0
1	21.0–28.0	23.0
2	24.0–32.0	27.0
3	24.0–39.0	28.0
4	29.0 +	35.0
5	38.0 +	

Component I

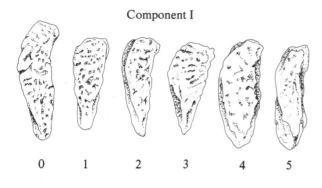

0	1	2	3	4	5

Component II

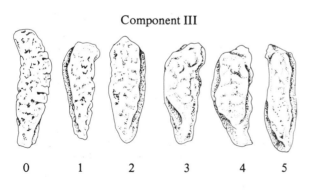

0	1	2	3	4	5

Component III

0	1	2	3	4	5

Figure 3.8 Medial surface of the symphysis pubis of females. All pubes depicted in Component I are on the left side; those in Component II are 0, 2, 4, right side, and 1, 3, 5, left side. Component III pubes are 0, 2, 5, right side, and 1, 3, 4, left side. Component II is tilted laterally to emphasize the ventural aspect (after Gilbert & McKern).

three age estimates, it follows that the individual was probably 19+ at death.

It is not advisable to use symphyses that are in any way eroded or show noticeably abnormal form. Moreover, it is important to note that these age-changes have so far only been well worked out for males (Stewart, 1957), although Gilbert & McKern (1973) provide a tentative scheme for females (Fig. 3.8, Table 1b) which can be used provided it is understood that childbirth can modify pubic morphology to some extent. Also, we still need far more information on changes at the pubic symphysis in various world populations. This includes work on pubic age data on ancient series, as for instance being explored in Japan (Kobayashi, 1964).

Table 1b
Age limits of the component stages

Stage	No. cases	Age range	Mean
Component I			
0	3	14–24	18.00
1	23	13–25	20.04
2	27	18–40	29.81
3	16	22–40	31.00
4	27	28–59	40.80
5	15	33–59	48.00
Component II			
0	11	13–22	18.63
1	21	16–40	22.52
2	25	18–40	29.64
3	19	27–57	38.77
4	28	21–58	40.90
5	9	36–59	48.50
Component III			
0	30	13–25	20.23
1	8	18–34	25.75
2	14	22–40	32.00
3	21	22–57	35.60
4	27	21–58	39.90
5	11	36–59	49.40

3   Dental attrition

Attrition may be defined as the wearing away of tooth substance during mastication by the rubbing of one tooth surface against another, together with the abrasive effect of any hard material present in the food (Campbell, 1939). Although this is mainly restricted to the occlusal surfaces, owing to slight movement, shallow facets may be produced where two adjacent teeth touch. To what extent attrition in earlier and in modern primitive man is the result of abrasives in the food, as opposed to the hard and fibrous nature of their foods needing much and strong mastication, is still debated (Dobrovsky, 1946). Suffice it to say here that most normal teeth show some degree of wear, but this is far less marked in recent civilized groups than in ancient and modern primitive populations. The degree of wear is dictated principally by the amount and strength of mastication as well as by accidental abrasives in the food. Cultural practices also result in some (usually restricted) degree of wear; Pendersen (1952), for example, notes that hide-chewing by Eskimo women produces abnormal wear on the anterior teeth.

Study of the degree of attrition has three principal uses when dealing with excavated material. First and foremost, it can assist in age determination (unless the antiquity and provenance are unknown). Secondly, it may help to determine the numbers of individuals present, provided that there are sufficient differences in the degrees of wear. Thirdly, similarity in dental wear may enable fragments of jaw to be associated with the same individual, especially if there is little other evidence available. Incidentally, atypical wear (produced by the hoaxer) was one of the clues that helped to expose the Piltdown fraud (Weiner *et al.*, 1953).

Although in using attrition to determine the age of individuals one can be fairly certain that there is a continual increase as the person gets older, there are a number of complicating factors to be borne in mind. In the first place, there may be a slight difference in dental wear between the sexes, for cultural and physical reasons; although such differences do not appear to be great enough to upset age estimates. With the eruption of the second and then the third molars, it is possible that the rate of wear may decrease slightly, but here again, the evidence of numerous early British specimens suggests that any slowing up is fairly insignificant. The least important factor is senility and the consequent reduction in chewing vigour. In Britain, as in other parts of the world, the average adult life expectancy until only a few centuries ago, was about 30–35 years, so that few aged individuals are to be expected in early populations.

Zuhrt (1955) has made use of attrition in assessing the ages of individuals in 8th–14th century burial material from Germany. In the immature specimens, he noted the degree of attrition on the first molar at the time of the second molar eruption (usually about 6 years after the first); also the degree of attrition on these two at the time of third molar eruption (normally about 6–10 years after the other two). He was thus able to establish roughly how much attrition took place over a fixed period of years. This information was then applied in ageing the adults. Miles (1963) undertook the same kind of dental ageing on early British material and this method has been applied satisfactorily to other archaeological samples elsewhere (Nowell, 1978).

Ideally, this process of checking attrition rates in immature specimens should be undertaken whenever skeletal series are large enough. Clearly this is not possible in the study of recent civilized groups, where soft foods result in little wear, but the material of most other populations should yield information in this respect. In addition, the degree of wear should be checked when possible, against the age-range given by the pubic symphysis (at least, in males). These factors

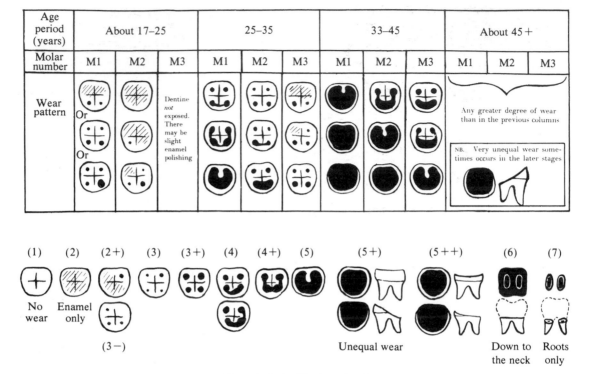

Numerical classification of molar wear

(N.B. Some patterns are more common than others, and there are minor differences between upper and lower dentitions.)

Figure 3.9   A tentative classification of age in Neolithic to Medieval British skulls, based on molar wear.

enable age-attrition standards to be established for a group, and should greatly assist in determining the age of other specimens where the teeth are the only available criteria.

Theoretically, one should not assess the age of specimens on attrition standards established on the basis of material belonging to another archaeological period and to a different area. However, it is fortunate that the rates of wear in earlier British populations do not appear to have changed much from Neolithic to Medieval times, and the attrition chart prepared for this handbook (Fig. 3.9) should be roughly correct for all these periods. It is based on adults and children of several early British groups, and in one series (Maiden Castle Iron Age), the degrees of wear have been checked against the age given by the pubic symphyseal face. Thus the estimates should in most cases reliably place the individual in a particular age range.

4   Microscopic methods of age determination

Following initial studies by Kerley (1965), there has been growing interest in the possibility of ageing bone by changes at a histological level (numbers of osteones, osteone fragments and non-Haversian canales). The estimation of the percentage of lamellar bone in four selected 100 power fields was made from outer third of the cortex in ground sections from the mid-shaft of the femur, tibia and fibula. Age was estimated using regression formulae established in relation to the histological changes noted.

Further investigation of this technique (Ortner, 1975) suggests that there are methodological hazards in using this method of ageing, and there is the possibility that the nature of physical activity in an individual's life, variations in diet, and endocrine status, could all influence this histological picture. Criticisms of the method have, however, led to further studies (Bouvier & Ubelaker, 1977; Kerley & Ubelaker, 1978) which have tended to confirm the value of this technique.

5   Palaeodemography

During the past ten years, there has been growing interest in what might be called the vital statistics of past populations. There are in fact three primary areas of human demography that can be considered in relation to earlier peoples: *a)* population growth and decline; *b)* the composition of communities; *c)* the distribution of populations in space and time (Swedlund and Armelagos, 1976). This information is similar in some respects to the information collected by demographers studying modern groups, but there are differences, determined by the fact that in archaeology we are dealing with unnamed and to some extent unknown individuals (Brothwell, 1971). Nevertheless, we are able to consider such demographic parameters as age-group composition, sex ratios and the reasons for mortality in a community, but only as a result of considerable detective work in relation to the human remains. Procedures for ageing and sexing have already been given, and it is now useful to consider how this can be treated at a population level. Sometimes, of course, there is a 'bonus' of information in the form of tombstone information, but this is uncommon for most earlier cultures.

Studies in palaeodemography are mainly in the form of regional surveys (Johnston &

Snow, 1961; Angel, 1969; Acsádi & Nemeskéri, 1970; Brothwell, 1972; Masset, 1973; Ubelaker, 1974; Ward & Weiss, 1976). In viewing such literature, one can see a number of general points emerging.

1) Life expectancy was variable from group to group, but was generally far less in the past than in modern advanced societies.

2) Trends may be seen through time, related to diet, disease and other aspects of the culture and environment.

   It should be noted here that estimates of the numbers of children born to individual females have been estimated on the development of 'parturition scars' near the pubic symphysis (Putschar 1976). Before it can be safely applied to a range of archaeological material, further studies are needed (Holt, 1978).

3) Differences occur between males and females in such variables as age at death.

4) Age-group composition can show variation from sample to sample and this can be at times related to cultural factors. In Fig. 3.10 Greek, Egyptian and Nubian children and sub-adults show much variation, only the Lerna sample being 'natural' in showing a high infant mortality.

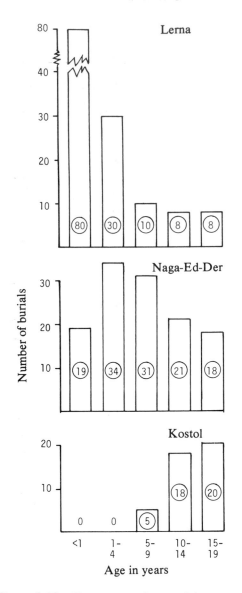

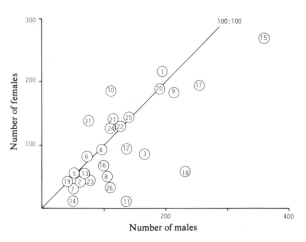

Figure 3.11   Number of adult females plotted against the number of adult males for thirty-one early series (from Brothwell, 1971).

Figure 3.10   Frequency of non-adults in various age groups, from Lerna in Greece, Naga-Ed-Der in Egypt, and Kostol in Nubia.

5) Estimates of the total population of a settlement is not easy from burial samples alone, and is best produced from

additional information on quantity of finds of the material culture, including evidence of the density of habitations.

6) Sex ratios should be approximately 1:1. However, there may be marked variation in archaeological samples, and this is likely to indicate the influence of cultural factors that can produce bias (Fig. 3.11).

Palaeodemographic studies also offer an opportunity to attempt to link together various biological information on skeletons, not only of age, sex and disease, but also metric and non-metric data, since movements of peoples will to some extent leave their mark in morphological changes through time.

As regards the presentation of demographic information, the extent to which it is worth constructing life tables of the kind seen in Table 2 depends very much on the size of the sample. A simpler tabulation may be sufficient unless special emphasis is being given to palaeodemography.

Table 2

Standard Life Table

A)   With 10 infants

Age	d (x)	d'(x)	l (x)	q (x)	L (x)	e (x)
0	10	3.92	100.00	.039	98.0	21.8
1	15	5.88	96.07	.061	93.1	21.6
2	44	17.25	90.19	.191	467.8	22.0
7	17	6.66	72.94	.091	278.4	21.6
11	8	3.13	66.27	.047	323.5	19.6
16	7	2.74	63.13	.043	308.8	15.5
21	50	19.60	60.39	.325	252.9	11.1
26	33	12.94	40.78	.317	171.5	10.2
31	25	9.80	27.84	.352	114.7	8.8
36	15	5.88	18.03	.326	75.4	7.2
41	19	7.45	12.15	.613	42.1	4.5
46	11	4.31	4.70	.917	12.7	2.9
51	1	.39	.39	1.000	.9	2.5

B)   With 160 infants

Age	d'(x)	d (x)	l (x)	q (x)	L (x)	e (x)
0	160	39.50	100.00	.395	80.2	13.9
1	15	3.70	60.49	.061	58.6	21.6
2	44	10.86	56.79	.191	256.7	22.0
7	17	4.19	45.92	.091	175.3	21.6
11	8	1.97	41.72	.047	203.7	19.6
16	7	1.72	39.75	.043	194.4	15.5
21	50	12.34	38.02	.325	159.2	11.1
26	33	8.14	25.67	.317	108.0	10.2
31	25	6.17	17.53	.352	72.2	8.8
36	15	3.70	11.35	.326	47.5	7.2
41	19	4.69	7.65	.613	26.5	4.5
46	11	2.71	2.96	.917	8.0	2.9
51	1	.24	.24	1.000	.6	2.5

Standard life tables calculated for an early Sudanese cemetery at Meinarti (Swedlund & Armelagos, 1976). In Section *A* the original data is presented, while in *B* the infant component was increased to 40 per cent, and probably gives more reliable figures for this ancient population as a result. Note that column d(x) represents the number of individuals found at different age levels, while l(x) gives a percentage survival into the next age category down the column.

IV Measurement and morphological analysis of human bones

1 Measuring the bones

In both preliminary and final reports it is usual to find a relatively large number of skull measurements, but only a few post-cranial dimensions, and these generally restricted to long-bones. There are a number of reasons for this. First of all, the skull has always excited considerable interest owing to its close association with man's brain. Secondly, differences in head features tend to be more easily recognizable than those in other parts of the body. Thirdly, the composite nature of the skull and its form enable the definition of fairly concise points from which measurements could be taken. Possibly another factor has been the tendency for the archaeologist or curator to retain only the skull for museum collections.

In recent years, the use of bone measurements, as undertaken by earlier anthropologists, has come under criticism. Clearly we can no longer rely on one or two metrical features, such as skull length and breadth, to differentiate accurately the various modern and ancient ethnic groups of the world. However, to suggest that craniometry, or for that matter any metrical work, is out of date (Boyd, 1950) is going too far. The study of earlier peoples, with the exception of mummified remains, will always depend upon bone studies. Whatever contribution chemical analyses may make to the study of skeletal remains, much of the comparison between individuals and groups will rest on recording morphological differences. This is not to say that in the interpretation of results we must not be far more cautious than our predecessors. We know, for example, that although the growth, size and form of a bone are to some extent genetically controlled, such factors are also altered by environment (Hiernaux, 1963). This plasticity of the human form is particularly evident from various studies on immigrants (Shapiro, 1939; Lasker, 1946; Goldstein, 1943), and twins. Also, even in comparatively small collections of skeletal material, there may be considerable variability in measurement or general form within a group, resulting sometimes in an overlap between the normal distribution of two or more groups (as in the British Neolithic and Bronze Age populations).

It is evident that while examining skeletons these factors must be borne in mind, and older methods examined critically for their usefulness. Statistical procedures are now helping far more than in the past, and the old indices are being replaced by multifactorial analysis and significance tests. Even other methods, such as cephalometric radiography may be worth considering in some special investigations (Savara, 1965). It is possible that as our knowledge of the genetics of human bone growth increases, modifications in the present measurements may be necessary. Other features, such as the height of the temporal lines or the size of the coronoid process, must be interpreted not as the result of an inherent tendency so much as the result of muscular and environmental influences. These considerations are equally important in reconstructing the physique of earlier forms of man.

The problem as to the number of measurements that the archaeologist may be expected to take in preparing a report, is one on which the opinions of anthropologists will differ. In the case of the skull, it has been thought advisable to give two lists; the first being the measurements most commonly used in general reports, the second, some of those less frequently employed.

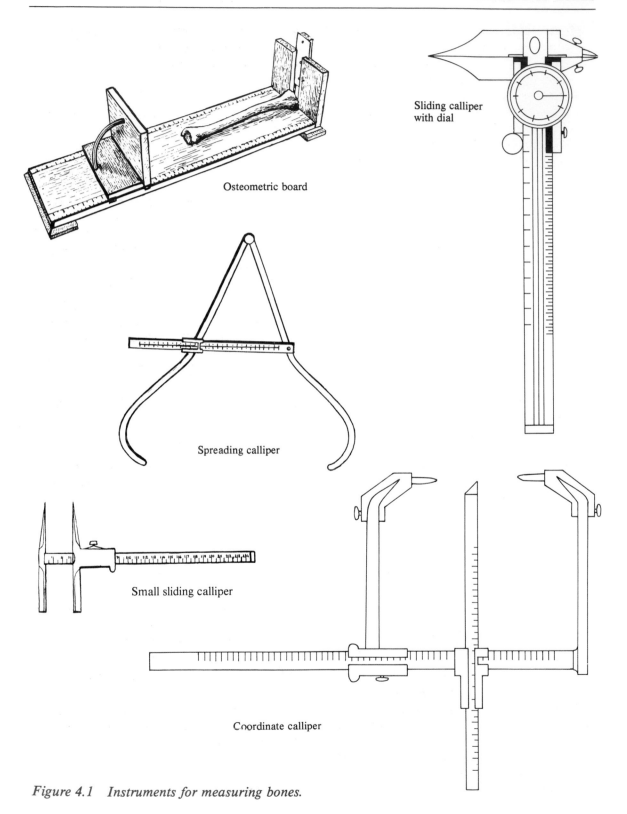

Osteometric board

Sliding calliper
with dial

Spreading calliper

Small sliding calliper

Coordinate calliper

Figure 4.1   Instruments for measuring bones.

Useful reviews of osteometric points and measurements include the works of Buxton & Morant (1933), Stewart (1947), and Ashley Montagu (1951). In some studies, specialized additional measurements have been taken (Howells, 1973), and even the small ear ossicles have been evaluated metrically (Masali, 1964). In the following brief list, no measurements involving the use of the Frankfurt horizontal plane have been used, for this depends upon points which themselves may vary in position relative to other parts of the skull. No angles are included, either because of their dependence upon this horizontal plane, or because their measurement requires specialized equipment. Facial subtences, although sometimes of value in deciding upon the degree of 'flatness', are beyond the scope of preliminary reporting. Indices, the 'percental relation of two measurements', are considered separately (p.87).

It cannot be too firmly emphasized that when possible definitions of all measurements that have been taken should be given in the report, and all care must be taken in following defined methods. Measurements of young and adult skulls and both sexes should be kept separate. Some practice is usually necessary before complete accuracy is possible, and before the worker feels at ease with the instruments. Those most commonly used are as follows (Fig. 4.1).

i) Spreading calliper. If this is not available, an ordinary engineer's calliper without a gradation scale can be used, being checked against a 30 cm rule.

ii) Sliding calliper (25 cm). It should be noted that in these days of 'computerized osteometry', calipers can be purchased which automate all stages of data readout and reduction (Garn & Helmrich, 1967). However, this is expensive equipment which is still not easy to justify in the budget of all laboratories.

iii) Small sliding calliper (about 15 cm). This is useful for small skull measurements. A newer version is the calliper with dial.

iv) Tape. Preferably a non-elastic linen tape or a narrow steel tape.

v) Mandible board. A simple form can easily be made.

vi) Osteometric board.

vii) Special studies may employ other equipment, such as the coordinate calliper (Howells, 1973).

2   Points or 'landmarks' of the skull

(Figs. 4.2–4.5)

The following are the 'landmarks' of the skull most commonly recognized, and used in recording the measurements of a skull.

Bregma, the point at which the coronal and sagittal sutures meet.

Lambda, the point at which the sagittal and lambdoid sutures meet.

Opisthion, the point at which the external and internal surfaces of the occipital bone meet on the posterior margin of the foramen magnum in its median plane.

Nasion, the mid-point of the suture between the frontal and the two nasal bones.

Basion, the lowest point on the external surface of the anterior margin of the foramen magnum in its median plane.

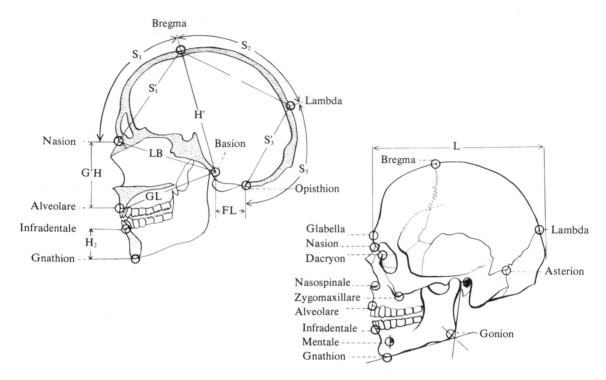

Figure 4.2   The principal craniometric points of the skull.

Glabella, the most prominent point between the supraciliary arches in the median sagittal plane.

Alveolare (alveolar point), the lowest point on the alveolar process between the sockets of the two central incisor teeth.

Asterion, the point at which the sutures between the temporal, parietal, and occipital bones meet.

Dacryon, the point at which the sutures between the frontal, maxillary and lacrimal bones meet.

Nariale, the lowest point on the inferior margin of the nasal aperture each side of the spine.

Zygomaxillare, the lowest point on the suture between the zygomatic and the maxillary bones.

Orale, the mid-point of a line tangential to the posterior margins of the sockets of the two upper central incisor teeth.

Staphylion, the point at which a line tangential to the two curves in the posterior border of the palate crosses the interpalatine suture.

Endomolare, the mid-point on the inner margin of the socket of the second upper molar tooth.

Nasospinale, a point in the median sagittal plane and situated on a line between both nariale; usually it is at the base of the nasal spine.

Gnathion, middle point on the lower border of the mandible.

Intradentale, the most antero-superior point on the alveolar margin between the lower central incisors in the lower jaw.

Gonion, the most lateral external point of junction of the mandibular body and ascending ramus.

Mentale, the most anterior point along the margin of the mental foramen of the lower jaw.

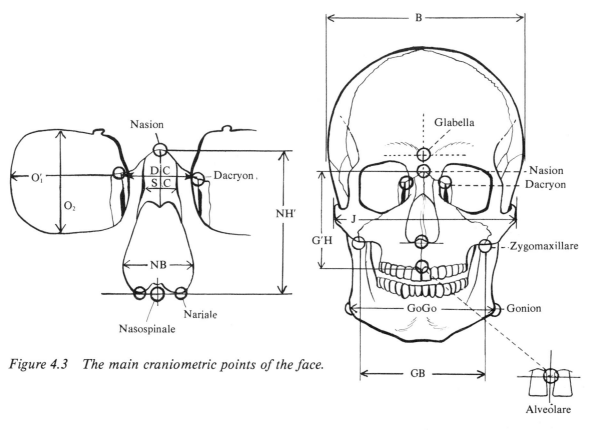

Figure 4.3   The main craniometric points of the face.

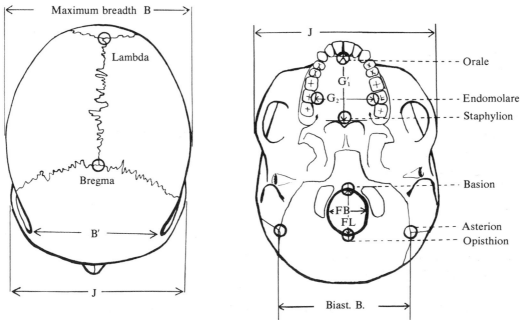

Figure 4.4   Other craniometric points of the vault and skull base.

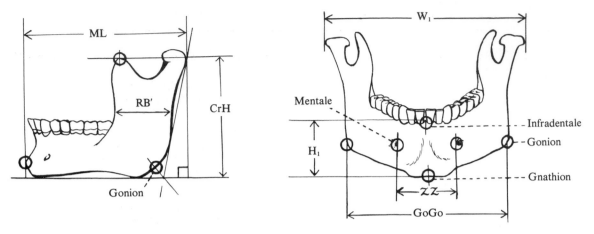

Figure 4.5   Craniometric points and major dimensions of the mandible.

3   Skull measurements

a) Measurements most commonly reported   (Figs 4.2–4.5)

Description of measurement	Biometric symbol*
I. Maximum cranial length. Greatest length in median sagittal plane from glabella to the most posterior point on the occipital. (Spreading calliper.)	L
II. Maximum Breadth. Greatest bi-parietal breadth, taken at right angles to the mid-sagittal plane. (Spreading calliper.)	B
III. Basi-bregmatic height. From the basion to the bregma. (Spreading calliper.)	H'
IV. Basi-nasal length. From the basion to the nasion. (Sliding calliper.)	LB
V. Basi-alveloar length. From the basion to the alveolare. (Sliding calliper.)	GL
VI. Upper facial height. From the nasion to the alveolare. (Sliding calliper.)	G'H
VII. Bimaxillary breadth. From one zygomaxillare to the other. (Small sliding calliper.)	GB

Description of measurement	Biometric symbol*
VIII. Bizygomatic breadth. Greatest breadth between zygomatic arches. (Spreading or sliding calliper.)	J
IX. Nasal height. From nasion to nasospinale. (Small sliding calliper.)	NH'
X. Nasal breadth. The maximum breadth of the nasal (pyriform) aperture between the anterior surfaces of its lateral margins; perpendicular to the mid-sagittal plane. (Small sliding calliper.)	NB
XI. Orbital breadth. Greatest breadth of the orbit measured from the dacryon to the anterior surface of its lateral margin. (Small sliding calliper.)	O'₁ (left or right)
XIII. Palatal length. From the staphylion to the orale. (Small sliding calliper.)	G'₁
XIV. Palatal breadth. From one endomolare to the other. (Small sliding calliper.)	G₂

* This is not the symbol used by Howells (1973), but is the British equivalent.

b) Measurements less frequently employed

Description of measurement	Biometric symbol*
I'. Minimum frontal breadth. Smallest diameter between the temporal crests on the frontal bone. (Spreading or sliding calliper.)	B'
II'. Frontal arc. Minimum distance from nasion to bregma taken over the surface of the frontal bone. (Tape.)	S_1
III'. Parietal arc. Surface distance from bregma to lambda. (Tape.)	S_2
IV'. Occipital arc. Surface distance from lambda to opisthion. (Tape.)	S_3
V'. Frontal chord. Minimum distance from nasion to bregma. (Sliding calliper.)	S'_1
VI'. Parietal chord. Minimum distance from bregma to lambda. (Sliding calliper.)	S'_2
VII'. Occipital chord. Minimum distance from lambda to opisthion. (Sliding calliper.)	S'_3
VIII'. Foraminal length. From the basion to opisthion. (Small sliding calliper.)	FL
IX'. Foraminal breadth. Maximum internal breadth of the foramen magnum. (Small sliding calliper.)	FB
X'. Simotic chord. Minimum breadth of the nasal bones, taken along the maxillo-nasal sutures. (Small sliding calliper.)	SC
XI'. Bi-dacryonic arc. Shortest surface distance from one dacryon to the other. (Tape.)	DA
XII'. Bi-dacryonic chord. Minimum distance from one dacryon to the other. (Small sliding calliper.)	DC
XIII'. Biasterionic breadth. The diameter from one asterion to the other. (Sliding calliper.)	Biast B
XIV'. Maximum horizontal perimeter above the superciliary arches and through the most projecting part of the occiput. (Tape.)	U

Description of measurement	Biometric symbol*
XV'. Transverse biporial arc. Measured from one porion, through bregma, to the other. (Tape).	BQ'
XVI'. Nasal height from nasion to left nariale. (Small sliding calliper). Some workers prefer this dimension to NH'.	NH
XVII'. Intercondylar (bicondylar) width. Diameter between most external points of the mandibular condyles. (Sliding calliper.)	W_1
XVIII'. Bigonial breadth. The distance between the gonia. (Sliding calliper.)	Go Go
XIX'. Foramen mentalia breadth. From one mentale to the other. (Small sliding calliper.)	ZZ
XX'. Minimum ramus breadth. Smallest distance between anterior and posterior borders of the ascending ramus. (Small sliding calliper.)	RB' (left or right)
XXI'. Symphysial height. Distance between gnathion and infradentale. (Small sliding calliper.)	H_1
XXII'. Maximum projective mandibular length. Distance between the most posterior points on the condyles to the most anterior point of the chin. (The lower jaw is placed on the mandible board, the condyles in contact with a vertical upright. Any rocking of the bone is prevented by pressing down on the two second molars. The other vertical upright is placed parallel to the first, and at the most anterior point on the chin; the distance is then recorded.)	ML
XXIII'. Coronoid height. Mandible placed on board and held steady as for ML. Maximum height of the coronoid process from the base. (Small sliding calliper or rule.)	CrH (left or right)
XXIV'. Cranial capacity.	C

4   Measuring cranial capacity

There are a number of practical and theoretical procedures for ascertaining the endocranial capacity of a skull. These have been reviewed by Hambly (1947) and Tildesley (1956).

In the recent major groups of mankind, differences of over 300 cm have been found for this measure. The direct method of estimate is usually to fill the cranial cavity (after sealing the foramina with cotton wool) with mustard seed or small shot. When completely filled to the foramen magnum, the material is emptied into a cubic centimetre measuring glass and the reading taken directly. Practice is needed in order to get a

similar degree of packing each time, but this does not eliminate differences that can occur between workers. Various formulae have been evolved for estimating the capacity, especially when the vault is not complete. These generally utilize the three dimensions L, B and H_1. However, although there are these formulae for the different major groups of mankind, it is possible that they do not sufficiently take into account the variability of form within a group, and this may be a source of some error. Only in special studies, especially of Palaeolithic remains, are capacity estimates of importance.

5   The measurement of teeth

It is over a hundred years since the variation in teeth began to be measured. In all, over twenty measurements have been defined, but as the more recent detailed studies by Lunt (1969), Wolpoff (1971) and Frayer (1978) demonstrate, fewer dimensions are usually selected. Crown dimensions have been especially important, and in particular the two measurements taken at right angles:

a) maximum mesiodistal diameter

b) labiolingual diameter (taken at right angles to the mesiodistal dimension)

These measurements are affected by severe dental attrition, when the degree of enamel loss may make the measurements unreliable.

Root lengths have not been taken so commonly. This length can be defined as the greatest length parallel the long axis of the tooth. A variant of this for the molars, is by measuring the longest root, from tooth neck to root apex. Whatever measurement is taken on teeth, it is particularly important to define precisely the methods used. In the case of roots in particular, the tooth surface may be clearly deformed, in which case it is best to exclude such defective areas.

Regarding the evolution of the hominids,

there is a growing literature on fossil teeth and dental morphological trends. It can be seen, for instance, in Figure 4.6, that in the hominid line, if estimated body weight is plotted with post-canine size, there is a very noticeable decrease in the latter (in noticeable contrast to pongids and the Plio-Pleistocene australopithecines. Gould, 1975).

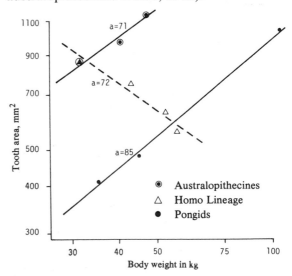

Figure 4.6   Scaling of postcanine tooth areas in great apes, australopithecines and hominines (after Gould, 1975).

6   Post-cranial measurements

i) Long-bones

The measurements that can be taken on the long-bones are also fairly numerous, and it has thus been thought advisable to divide them into two groups according to their importance (largely determined by their usefulness in determining stature). Most of the useful dimensions have been defined and employed in the study of the ancient inhabitants of Jebel Moya (Sudan) by Mukherjee, Rao & Trevor (1955). Scholfield (1959) reviews much of the earlier work on femoral variations. Specialized measurements, such as the angle of femoral torsion seem more likely to have general value in fossil hominid and growth studies (Vare, 1974) rather than in analyses of archaeological material in general.

It should also be noted that long-bone measurements are not simply of value in studying adults. As Merchant & Ubelaker (1977) demonstrated in their detailed study of children and infants from Amerindian cemeteries, limb dimensions can suggest depressed growth rates linked to environmental and possibly genetic factors (Fig. 4.7).

a) *Primary long-bone measurements* (See Fig. 4.8)

Femur. Maximum length (FeL$_1$), from the medial condyle at the distal end of the femur to the most proximal part of the head. In determining this, the posterior border (linea aspera side) should face downwards on the osteometric board.

Tibia. Total length (TiL$_1$), from the lateral condyle at the proximal end of the tibia to the tip of the medial malleolus. The posterior (back of the leg) surface should face downwards on the osteometric board, the long axis of the bone being parallel to the long axis of the board.

Humerus. Maximum length (HuL$_1$), from the medial margin of the trochlea at the distal end to the head of the bone. The head is placed against the fixed vertical of the board,

and the other upright to the distal extremity; the bone is then moved up and down as well as from side to side until a maximum length is obtained.

Radius. Maximum length (RaL$_1$), from the head to the tip of the styloid process at the distal end. The measuring procedure is the same as for the humerus.

Ulna. Maximum length (UlL$_1$), from the top of the olecranon to the tip of the styloid process at the distal end. Measuring procedure as with the humerus.

Fibula. Maximum length (FiL$_1$), between the proximal and distal extremities. Measuring procedure as with the humerus.

b) *Secondary long-bone measurements*

The following list is not exhaustive, but it gives the measurements now most usually adopted in more specialized reports.

Femur. Oblique length (FeL$_2$), taken from both distal condyles (placed against the fixed upright of the osteometric board) to the femoral head, posterior border facing downwards.

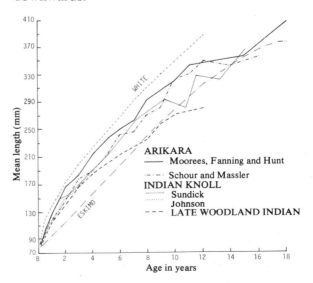

Figure 4.7 Comparison of cross-sectional growth curves in the femur of Arikara, Indian Knoll, Late Woodland Indian, Eskimo and Modern White populations (after Merchant & Ubelaker 1977).

Trochanteric length (FeL₃), taken from the lateral condyle to the tip of the greater trochanter. The long axis of the bone parallel to that of the osteometric board (or vertical bar, if an anthropometer is used).

Minimum antero-posterior diameter (FeD₁) of the shaft below the lesser trochanter, with the gluteal tuberosity avoided (using the sliding calliper).

Transverse diameter (FeD₂) of the shaft at same level as and perpendicular to FeD₁, taken with a sliding calliper.

Bicondylar breadth (FeE₁), taken at the distal extremity, with both condyles in contact with the graduated bar of a sliding calliper.

Tibia. Oblique length (TiL₂) from both proximal condyles to the tip of the medial malleolus, its posterior surface facing

The position of the tibia when recording the two dimensions at the nutrient foramen.

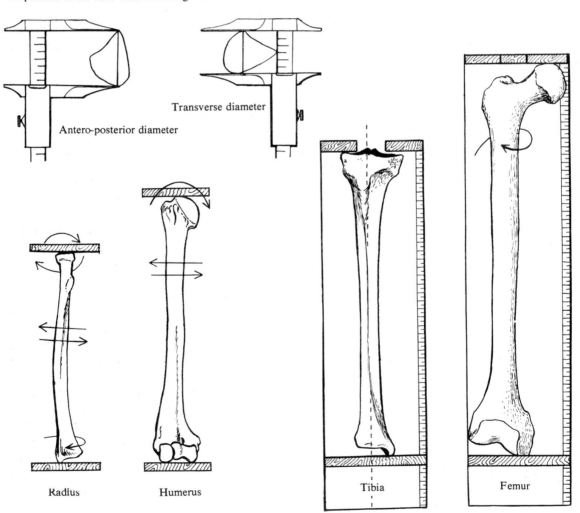

Antero-posterior diameter

Transverse diameter

Radius

Humerus

Tibia

Femur

Figure 4.8   The positioning of long-bones in order to record maximum lengths. Horizontal arrows denote movement from side to side. Curved arrows denote circular movement. The broken line shows the axis of the long-bone parallel to that of the osteometric board. Also the procedure for taking the upper tibial shaft dimensions.

downwards on the osteometric board.

Maximum antero-posterior diameter (TiD$_1$) of shaft at level of the nutrient foramen, measured with a sliding calliper.

Projective transverse diameter (TiD$_2$) of the shaft, measured with the anterior border of the bone touching the graduated bar of the sliding calliper equidistant from its arms.

Bicondylar breadth (TiE$_1$), or maximum medio-lateral breadth of the proximal extremity, taken with a sliding calliper.

Humerus. Maximum diameter (HuD$_1$) of the shaft equidistant from the terminals of HuL$_1$, measured with a sliding calliper.

Minimum diameter (HuD$_2$) of the shaft at same level as HuD$_1$, measured with a sliding calliper.

ii) Other post-cranial remains

Definitions of other skeletal measurements need not be given here. Suffice it to say that in specialized studies metrical data of other bones are recorded. Vallois (1946), and Olivier & Pineau (1957), for example, have undertaken studies on the scapula; Olivier (1956) has made a detailed analysis of clavicle dimensions; while special studies of smaller bones such as those of the ankle may also yield interesting information (Bostanci, 1959).

7   Skeletal indices

An index may be defined as the ratio of one measurement to another expressed as a percentage of the larger one. Well over twenty different indices have been formulated in the past by anthropologists, some being of far more use than others. Regarding the long-bones, the *platymeric* and *platycnemic indices* have received special attention, and are discussed separately. All indices, whether of the skull or other part of the skeleton, attempt to differentiate various bone shapes, and although somewhat rough and ready, they do show considerable variation in earlier populations. Perhaps the most well known index is the *cephalic index*, obtained by the formula:

$$\frac{\text{Maximum skull breadth} \times 100}{\text{Maximum skull length}}$$

This shows roughly the degree of round-headedness or long-headedness of an individual. The index range is usually between 65 and 90 and is conventionally divided into four sections as follows:

Dolichocephalic	= an index < 75.0
Mesocephalic	= 75.0–79.9
Brachycephalic	= 80.0–84.9
Hyperbrachycephalic	= 85 and over

This is not to say that a series of skulls will have indices falling into only one of these categories, but rather that in homogeneous groups, there will be a tendency for most indices to fall in one of these sections. The degree of variability to be found for this index in earlier populations is shown in Fig. 4.9. It is interesting that there have been noticeable changes in the mean cephalic index in Britain from Neolithic times onwards, and possibly this is the reason why its interest has been over-emphasized by earlier British anthropologists. However, with the advent of statistical procedures that consider together far more than two measurements, as a means of assessing the forms of the bones, the cephalic index has been rendered of far less value.

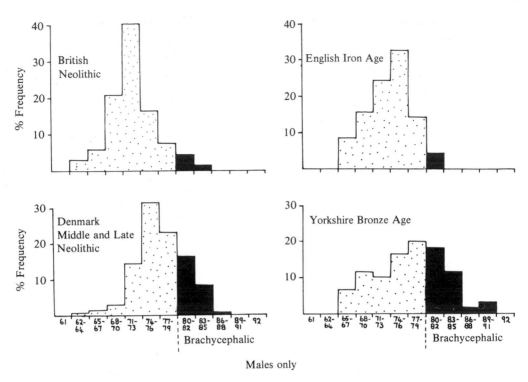

Males only

Figure 4.9   *Distribution of the cephalic (length–breadth) index in four prehistoric populations.*

8   Platymeria and platycnemia

In both the femur and tibia, the proximal part of the shaft sometimes shows noticeable differences in general shape between various populations. The shape in each case is recorded as the antero-posterior diameter (FeD_1 or TiD_1) and transverse diameter (FeD_2 or TiD_2) from which indices are calculated. In the case of the femur, the most noteworthy feature on the shaft is antero-posterior flattening, while in the tibia, it is the transverse flattening (Fig. 4.8).

i) Platymeria

This femur-shaft index is calculated by the following formula:

$$\frac{FeD_1 \times 100}{FeD_2}$$

and the results may range from below 70 to over 100. Two important divisions may be noted:

Platymeria   index below 84.9
Eurymeria   index 85 to 99.9

In Table 3 the so-called platymeric index is given for various modern and earlier populations to show the degree of variation that is encountered. Possible reasons for these differences have been reviewed by Townsley (1946), who considers that normal antero-posterior flattening of the neck part of the shaft is a mechanical adaptation, involving the economic use of material with sufficient strength to support body weight acting on the inclined femur neck. This author also noted that a low platymeric index may be associated with some pathological conditions such as osteoarthritis and osteoperiostitis. Other opinions include those of Cameron (1934), who rejected an earlier view that platymeria was due to squatting (see p.90), and suggested that it was caused by unwonted strain on the femora during childhood and early adolescence. Buxton (1938), on the other hand, considered that femoral shaft flattening took place when there was a shortage of bone material, believing that such bone inadequacy might have been caused by calcium or vitamin deficiency.

Table 3

Index of platymeria in fossil and recent man, showing the degree of variability so far known

Fossil man		
a) Cro-Magnon man	73	
b) Neanderthal man	77	
Recent groups		
a) Turks	73	
b) American Indians	74	
c) Andamanese	78	
d) Eskimo	81	
e) Australians	82	
f) English (17th C.)	85	

Various authors have claimed that platymeria is more common in females than in males. There is also a tendency for it to be more pronounced in the left femur than the right.

ii) Platycnemia

This tibia-shaft index is calculated from two dimensions taken at the level of the nutrient foramen, and may be defined thus:

$$\frac{TiD_2 \times 100}{TiD_1}$$

The index in man can be lower than 55 or higher than 70.

The two most important categories of the possible index range are:

Platycnemia   up to 62.9
Mesocnemia   63 to 69.9

It is important to take the measurement exactly at the distal edge of the nutrient foramen and not at any other part of the shaft, for as Vallois (1938) has emphasized, divergence of technique has rendered data by some authors valueless.

Various explanations, invoking pathological and muscular factors, have been suggested to account for this transverse flattening (Lovejoy, Burstein & Heiple, 1976). Cameron (1934) considers that platycnemia results from the persistent adoption of the squatting attitude when at rest, with the retroversion of the upper end of the bone. Andermann (1976) is somewhat critical of the reliability of this index, in view of the fact that it relies on the positioning of the nutrient foramen.

As in the case of platymeria, this index occurs more frequently in modern primitive groups and earlier man. Platycnemia and platymeria are not necessarily found associated.

9   Squatting facets

In some reports on human remains, attention is drawn to certain articular areas of the lower limb-bones. The evidence suggests that some articular surfaces may be found to differ in form not only within groups but also as regards the frequency of these forms in various populations. It is claimed that these differences are particularly marked between primitive and civilized groups, where the occurrence of the squatting habit differs widely. The argument is that in modern primitive groups and some earlier populations, where there is a lack or scarcity of household furniture, the habit of squatting is normal; and the result may be the extension of certain articular surfaces or the development of new facets. However, it is important to note that none of these squatting facets is constant within a group, and as yet few frequencies have been recorded for earlier groups.

By far the most noted squatting facet is that on the lower end of the tibia (Fig. 4.13), and appears to be frequent at least as early as Neolithic times. Further examples of such facets are to be found on the neck of the talus and at the femur, their antiquity extending back at least as far as the Neanderthalers (Trinkaus, 1975). Other changes may also take place. Martin (1932), for example, found that in the femora of squatting people there is generally a deeper intercondylar fossa, and the lateral edge of the patellar surface is often rounded.

The question as to whether such features indicate squatting clearly needs further investigation (Satinoff, 1972), and indeed it seems more likely that in the future such variation will be considered purely as non-metrical variation of a non-environmentally-induced kind.

10   The value of non-metrical data

Probably one of the most important criticisms that can be levelled against earlier anthropologists is that practically their whole attention was given to the measurement of bones. It is becoming increasingly obvious that one field which offers considerable promise is that of the study of non-metrical characters. In the past, this term has been used to refer to any morphological feature, such as mastoid size, shape of the nasal aperture, general facial shape, chin shape; but such observations seem best placed under the separate heading of 'anthroposcopic data'. This leaves out of account one type of feature, which is in most cases either clearly present or absent, and which may be grouped as *non-metrical*, or *discontinuous morphological* traits

As with metrical differences, it is still far from certain whether variations in the frequencies of discrete traits are controlled by one or many genes, although recent work suggests that at least some traits may have a fairly simple genetic background. Neither are we certain to what extent their frequency may be modified by the environment (Trinkhaus, 1978), or the degree of influence of other genes upon the expression of a particular character (Berry, 1975; and Saunders & Popovich, 1978).

However, the fact that some of the traits show considerable differences in frequency enables them to be employed in order to establish the 'distance' of one group from another. Of course, as with blood-group frequencies, because a character occurs in a high percentage of, say, the Eskimos and Chinese, it cannot be said on this evidence alone that the groups in question are closely related. Clearly it is necessary to take other factors into consideration, and particularly any definite knowledge of group relationships.

Although all the disçontinuous traits mentioned in the literature cannot be described here, some idea of the variety of traits may be given. It sometimes happens that the feature, when present, is always the same, or it may differ in size and shape or number; but usually the number of specimens available prevents these variations from receiving satisfactory analysis. Age, and to a lesser extent sex, should be noted in the analysis of such data (especially in the case of sutural features, where obliteration may make invalid any comparison between two widely divergent age-groups).

i) General review

Dixon (1900) discussed channels on the external surface of the frontal corresponding in position to the branches of the supraorbital nerves. Their occurrence, he found, varied from hardly any in Australians to over 50 per cent in Negroes.

Stallworthy (1932) noted the various forms of parietal foramina, and showed that percentage differences in three series amounted to 16 per cent.

Sullivan (1922) gave ethnic data on various anatomical features including the fifth cusp on the lower second molar, the tympanic perforation, and inca bones. The inca bone, detectable even in embryonic stages (Aichel 1915), is particularly well known for its frequency variations. The forms and aetiology of this bone were discussed by Hepburn (1908). Its frequency of occurrence varies between 0 per cent in Bavarian and Papuan groups to over 12 per cent among coastal Peruvians.

The non-metrical studies by Wood Jones (1931) continued by Wunderly (1939)

included work on the supraorbital foramina, infraorbital foramina and other anatomical points. Few frequencies, however, are yet available. Steida (1894) and Woo (1949) examined the direction and type of the transverse palatine suture. In the case of the anteriorly directed suture the percentage frequencies varied from 48 to 90 per cent. Laughlin & Jørgensen (1956) employed two new non-metrical features: a) the presence of the mylo-hyoid arch; and b) the direction of the superior sagittal sinus.

Riesenfeld (1956) has studied certain human foramina. For the incidence of multiple infraorbital foramina, he found percentage differences of up to 27 per cent, and for ethmoidal foramina, differences of 24 per cent.

Marshall (1955) and Broman (1957) considered the incidence of precondylar tubercles, obtaining frequencies ranging from 4 to 20 per cent.

In the case of the human dentition, there is a considerable literature on the variation of such features as molar pattern, extra teeth and congenital absence of teeth. Dahlberg (1951), for instance, gives for the congenital absence of one or more third molars differences varying between 2.6 per cent in West Africans to 50 per cent among the Sioux Indians.

ii) The range of traits

Discontinuous morphological traits have already been used with success by Laughlin & Jørgensen (1956) in analysing cranial series of the Greenland Eskimos. Following this, ten such characters were employed in order to ascertain their value in differentiating larger groups of mankind (Brothwell, 1959a). After applying 'distance' statistics, it was found that the non-metrical characters separated the groups as efficiently as a series of cranial measurements. The populations examined and percentage frequencies of the traits are given in Table 4. It may be noted

Table 4

Percentage frequencies of ten discontinuous morphological traits in fourteen populations §

(Male and female data have been combined)

	with tori mandibularis	with torus palatinus	with tori auditivi	with a metopic suture	showing fronto-temporal articulation	displaying wormian bones	of pteria with epipteric bones	of sides with parietal notch bones	showing orbital osteoporosis	of sides with multiple mental foramina
						Percentage				
1. Eskimo	39.81†	31.39†	0.20‡	0.28*	1.92*	25.00*	4.85*	14.62*	2.02*	2.22
2. Chinese	9.43†	3.59*	0.28*	8.17*	3.69*	80.32	11.28*	31.88	13.38*	11.43
3. Australian	6.25*	18.84†	14.47*	0.63*	23.85*	72.58	7.59*	12.86	5.26*	11.25
4. Melanesian	0.00†	0.00†	6.10*	2.02‡	15.07†	64.15	10.91†	19.29	1.65‡	9.71‡
5. Polynesian	4.05‡	24.42†	25.74‡	1.33†	2.29†	29.92*	9.09†	3.45	6.49‡	12.50‡
6. African Negro	14.15*	0.00‡	1.71‡	1.23†	13.55*	45.05*	11.58*	22.78	32.07*	8.01‡
7. N. American Indian	13.34‡	8.05†	12.76‡	1.45†	0.49	28.18*	2.28†	11.82	11.51*	3.65‡
8. Ancient Egyptian	2.41‡	1.33	1.98‡	3.87‡	1.55*	55.56*	10.08*	16.05	7.47‡	4.59*
9. Lachish	25.00	3.08	1.98	8.81‡	1.87‡	63.41	20.78	21.69	12.16	6.85
10. Anglo-Saxon	27.27	9.20*	0.00	8.30*	1.03	55.56	11.29	8.59	27.64	11.18
11. Iron Age Romano-British	37.23	9.71	1.82	9.91	2.73	71.03	19.89	36.04	35.00	9.18
12. Peruvian	1.79‡	19.31†	14.49†	2.56†	1.48*	51.85*	2.68*	20.45	8.09‡	10.00
13. German	0.00‡	4.87‡	5.00	8.37†	1.61†	75.00	18.29	12.5	6.47†	8.62
14. London (17th C.)	19.61	10.78*	5.63	9.09*	2.88*	36.02*	13.10*	8.08*	15.38	9.46

§   Certain published data have been omitted owing to lack of information as to the exact number of sides examined.

* = Data by the author and from the literature

† = Means calculated from various published sources

‡ = Already published data by one author

No mark = Original data by the author

that the populations selected included major and minor groupings, the results suggesting that these traits will be mainly of use in comparing peoples *within* major racial stocks (e.g. Anglo-Saxons and Germans) rather than in comparing, say, Australoids and Mongoloids.

It seems worth considering these traits used and others in some detail (Fig. 4.10), especially as it will help to give some idea of the extent of our knowledge concerning each one.

1) *Metopism*, the retention of the medio-frontal suture, may be discussed first. This suture usually disappears within the first one or two years after birth, but in some individuals it is distinguishable into late adult life. There is some evidence to suggest that its presence may depend upon a fairly simple genetic background. Ashley Montagu (1937) postulates genes for 'metopism' and 'non-metopism', while Torgersen (1951) considers it to be a dominant trait. Whatever the exact reason, metopism does show considerable geographical variation. As Sullivan (1922) has shown clearly among the Bolivians, even within a comparatively small region wide variation in incidence may occur.

2) *The pterion* is the area on the external surface of the skull where parts of the frontal, parietal, sphenoid and temporal bones meet. There are two main forms of articulation (Fig. 2.25), the two sides of the skull showing either the same or different patterns. The frequencies of the two forms have been well discussed by Ashley Montagu (1933), Murphy (1956) and others. Reasons for this dichotomy of form are clearly associated with ossification in this region, probably dependent upon both mechanical and genetic factors. Age, sex and the side (that is right or left) appear to have no influence on the type of contact.

3) *The mental foramina* are two elliptical holes on the outside of the mandible situated about 12 mm below the second premolar on each side. Multiple mental foramina (more than one on each side) are to be found in some primates occasionally including man. Although the numerical variability of the mental foramina is less than that of the ethmoidal and infraorbital foramina (Riesenfeld, 1956), there is at present far more data on this than on either of the other two. The reason for an increase in foramina is still unknown.

4) *Wormian bones* are extra (sutural) bones of the skull. The frequency of ossicles at the coronal or sagittal sutures has rarely been noted, and thus the term will here be restricted to ossicles occurring along the lambdoid suture. This does not include inca bones (Kadanoff & Mutafov, 1968) as these appear to depend upon different aetiological factors. Hess (1946) considers that the formation of wormian bones is related to a metabolic disorder of the mesoderm, while Torgersen (1954) claims that they are inherited as dominant traits with a penetrance of about 50 per cent. The expression of these genes, he says, depends upon modifiers influencing head shape generally. This view is supported by the work of El-Najjar & Dawson (1977). Bennett (1965), on the other hand suggests that these bones are determined purely by environmental (stress) factors.

5) *The epipteric bone* (or bones) found at pterion (Fig. 4.10) has also been reviewed by Ashley Montagu (1933) and Murphy (1956). Although these bones may be of various shapes and sizes, it is usual to give an overall frequency. Both Torgersen (1954) and Murphy (1956) consider that the occurrence of these bones is genetically controlled, although the evidence is not as yet great.

6) *The parietal notch bone* (or bones) found in the narrowest area of the incisura parietalis (Fig. 4.10) has so far received little attention. Laughlin & Jørgensen (1956) have found it of value in studying Eskimo isolates, and point out that prior to their study it had only been employed by Akabori (1933) on Japanese crania and by Weidenreich in his work on *'Sinanthropus'*. As yet, nothing is known of its aetiology.

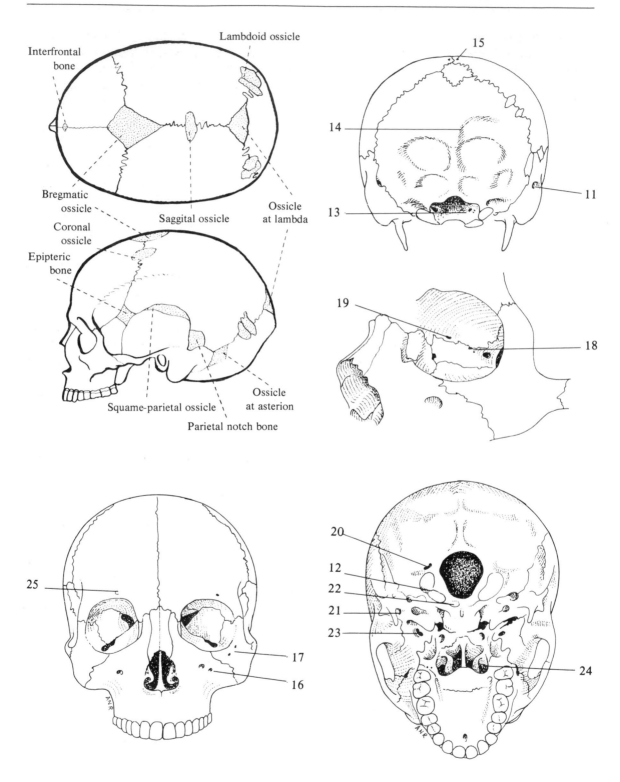

Figure 4.10   Sites of accessory (wormian) bones of the skull and other non-metrical traits.

Four tori (excluding the supraorbital and occipital tori) are found in the skull (Fig. 4.11). There is a considerable literature on these, which has been reviewed by Hrdlička (1940), Broek (1943), and Couturier (1959).

7) *The tori mandibulares* are bony protuberances on the lingual (inner) surface of the lower jaw occurring bilaterally, and usually restricted to the molar and premolar regions. They consist of a compact osseous tissue with a restricted number of Haversian canals (Broek). Their size may differ considerably, but they are usually sufficiently well defined for one to be certain of their presence. Earlier workers considered that these bumps arose as a result of functional factors, particularly chewing pressures. However, more recent work suggests that the tendency to develop this feature may be hereditary (Moorrees *et al.*, 1952; Lasker, 1950). The occurrence of this torus in children is particularly interesting; Moorrees (1957) found it in 24 per cent of the Aleut children that he studied.

8) *Torus palatinus* is a bony cigar-shaped prominence along the median line of the hard palate (Vidić, 1968), and is usually thought to be a post-natal hyperostosis (Miller & Roth, 1950). Factors that were at one time claimed to influence its development include mechanical stress, diet, disease and inheritance. There would now appear to be a growing concensus of opinion that the torus is an abnormality controlled by a specific gene or genes (Klatsky, 1956; Lasker, 1947).

9) *Tori auditivi,* the exostoses in the external auditory canal, are generally found on the posterior wall. They vary from compact bony tissue with some irregular Haversian canals to spongy-centred protuberances (Broek, 1943). The size of these tori may vary from small corrugations to large prominences almost filling the meatus. This anomaly should not be confused with the complete absence of the meatus

(Hrdlička, 1935; Risdon, 1939) which seems to be congenital. Mechanical irritation was the explanation first offered for these tori, but more recently the suggestion has been made that it is associated with an hereditary neurovascular derangement.

10) *Tori maxillares* have received less attention than the other tori. They are situated lingually in the region of the upper molars, and consist of very compact bony tissue. Although a number of percentage frequencies have been recorded, ranging from 2.5 to 17 per cent, the data are insufficient to warrant further discussion.

There is no need to provide a detailed list of all the other cranial traits that have now been used in analyses. Some are clearly more useful than others, and individuals may make their own choice based on the time available for recording all these facts or the nature of the research. Thus it is sufficient to note them here for consideration. (See also Fig. 4.10.)

11) The exsutural position of the mastoid foramen.

12) Double condylar facet on the occipital.

13) Double anterior condylar canal (foramen magnum).

14) Presence of a high nuchal line.

15) Presence of parietal foramina.

16) Presence of accessory infraorbital foramen.

17) Presence of zygomatic-facial foramen.

18) Presence of posterior ethmoid foramen (orbit).

19) The exsutural position of the anterior ethmoid foramen.

20) Posterior condylar canal is patent.

21) Foramen of Huschke present.

22) Precondylar tubercle present.

23) Foramen ovale incomplete.

24) Accessory lesser palatine foramen present.

25) Supra-orbital foramen complete.

iii) More recent cranial studies

Since the earlier comparative studies of non-metrical traits were published, a variety of other studies has appeared. Examples of the use of multi-trait analyses to evaluate populations and establish biological distances from other communities can be seen in studies by

Berry & Berry (1967), Berry, Berry & Ucko (1967), Knip (1971), Czarnetzki (1971/1972), Rightmire (1972), Dodo (1974), Larnach (1974b), Vargas (1974), Berry (1974), Sjovold (1974), Ossenberg (1974).

Following the early use of the Penrose 'size' and 'shape' statistic for the analysis of these traits (Brothwell, 1959a), further

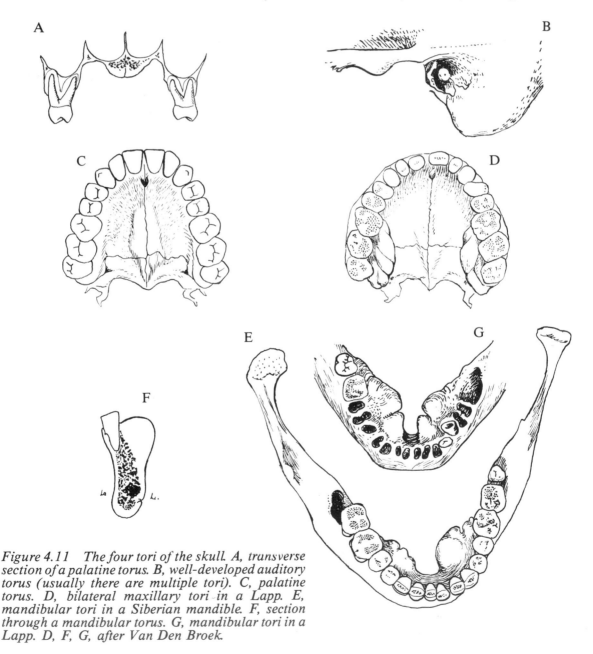

Figure 4.11   The four tori of the skull. A, transverse section of a palatine torus. B, well-developed auditory torus (usually there are multiple tori). C, palatine torus. D, bilateral maxillary tori in a Lapp. E, mandibular tori in a Siberian mandible. F, section through a mandibular torus. G, mandibular tori in a Lapp. D, F, G, after Van Den Broek.

techniques have been used. The 'mean measure of divergence' devised by C. A. B. Smith (Sjovold, 1977) has been especially employed, although some alternative possibilities have been considered (Green & Suchey, 1976; de Souza & Houghton, 1977; Finnegan & Cooprider, 1978).

iv) Dental traits

Developmentally-determined variation to be seen in teeth has been considered to some extent in other sections. However, it seems important to make the point here that such minor variants as shovelled upper incisors,

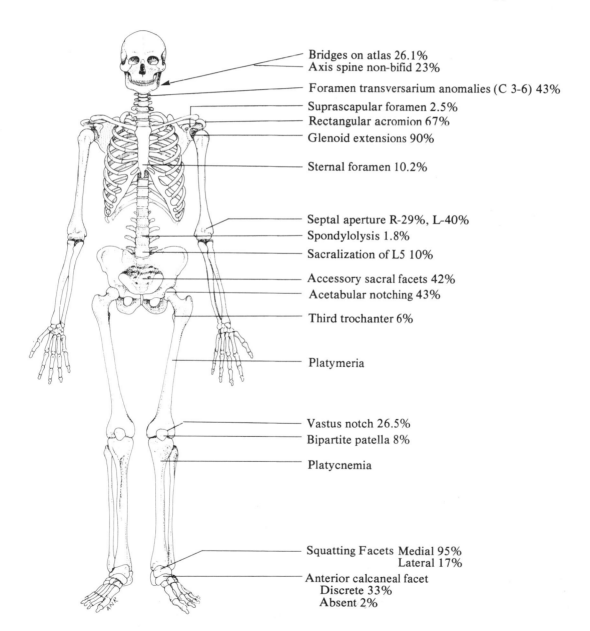

Bridges on atlas 26.1%
Axis spine non-bifid 23%

Foramen transversarium anomalies (C 3-6) 43%

Suprascapular foramen 2.5%
Rectangular acromion 67%
Glenoid extensions 90%

Sternal foramen 10.2%

Septal aperture R-29%, L-40%
Spondylolysis 1.8%
Sacralization of L5 10%

Accessory sacral facets 42%
Acetabular notching 43%

Third trochanter 6%

Platymeria

Vastus notch 26.5%
Bipartite patella 8%

Platycnemia

Squatting Facets Medial 95%
                Lateral 17%
Anterior calcaneal facet
    Discrete 33%
    Absent 2%

Figure 4.12   Infracranial variation in the skeletons from Fairty, Canada (Anderson, 1964).

Carabelli's cusp, an extra lingual cusp on PM₁, and so forth, can be used in the same way as other non-metrical traits. Examples of the kind of work that can be achieved, using such traits, are provided by Hanihara (1967), Hanihara, Tamada & Tanaka (1970), Berry (1974) and Baume and Crawford (1978).

iv) Non-metric variation in the post-cranial skeleton

Unlike the skull, bones of the post-cranial skeleton have as yet received little attention. Anderson (1963, 1968) and more recently Finnegan (1978) have shown the range of

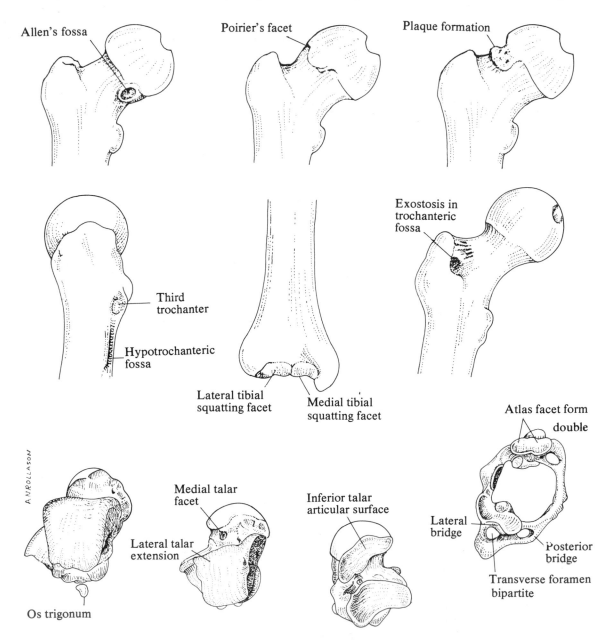

Figure 4.13   Variation in the post-cranial skeleton (after Finnegan, 1978).

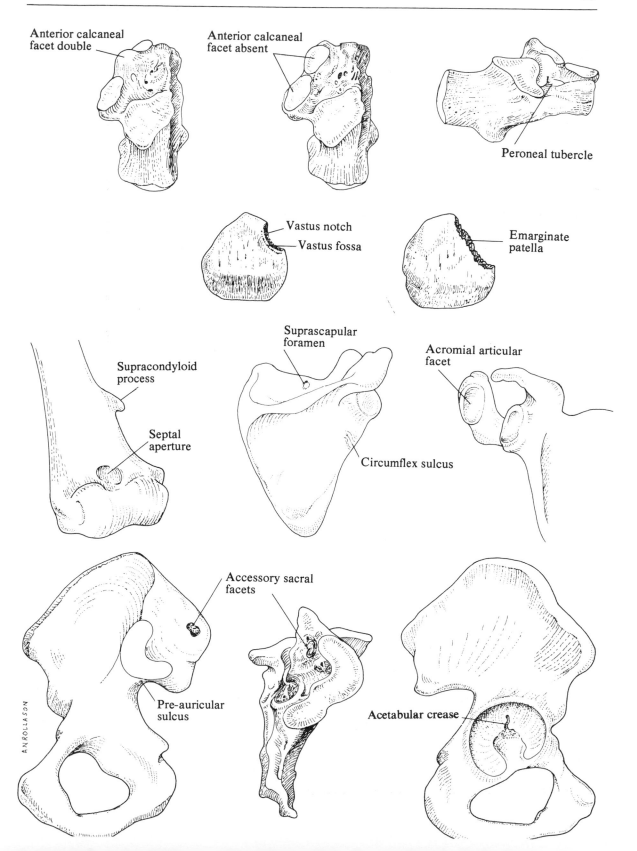

Anterior calcaneal facet double

Anterior calcaneal facet absent

Peroneal tubercle

Vastus notch
Vastus fossa

Emarginate patella

Supracondyloid process

Septal aperture

Suprascapular foramen

Circumflex sulcus

Acromial articular facet

Accessory sacral facets

Pre-auricular sulcus

Acetabular crease

ANROLLASON

variation which can occur and which seems worthy of study in earlier groups (Figs. 4.12 and 4.13). Investigations on regions of the skeleton still continue, of course, and may help to reveal further variation of comparative use (for example, Glanville, 1967; Tulsi, 1972).

11   Estimation of stature

Before methods of estimating stature in earlier groups are discussed, four basic anthropological tenets concerning height may be noted:

1. In a single individual, height increases until adulthood is reached, and decreases as senility approaches.

2. Within a population (whether homogeneous or mixed) considerable variation in stature will be found.

3. The mean statures of a number of populations may be widely different from each other, but when the general distribution of stature within each group is considered, a certain percentage of heights is found to be common to them all.

4. The average stature for females is smaller than that for males, whatever the group.

The height of a person is not easy to analyse genetically, and it is thought that many genes are involved in the control of body size and form. However, from the study of twins it has been deduced that over 90 per cent of a person's height is dictated by inheritance, whereas environmental factors, such as food and climate account for the other 10 per cent. Height studies both on modern and earlier populations must therefore be considered in terms of these two groups of factors.

Whereas in modern man overall height can be obtained directly, in dealing with skeletal material, estimates must be made after taking appropriate bone measurements. A rough estimate of height may be obtained by measuring the maximum length of the skeleton after articulating and making an allowance for joint cartilage and superficial soft tissues. However, as Harrison (1953) points out, this method is open to considerable error owing to the impossiblity of assessing the thickness of the cartilage and other structures. Estimates from the length of the articulated spinal column (Krogman, 1939) are probably also unreliable. The method of using the clavicle length has also been shown to be of no value (Jit & Singh, 1956), and Olivier & Pineau (1957) found that scapula dimensions gave inadequate results. On the other hand, estimation of adult stature from the length of metacarpal bones seems likely to give fairly acceptable heights (Musgrave & Harneja, 1978), and deserves consideration in the absence of major long bones.

As yet, the most reliable method of estimating stature is from the long-bones. If, for example, the skeletal remains of an individual included a complete femur, a maximum length could be obtained, and by the application of a special formula (known as a 'regression equation') the stature calculated. A number of such formulae have now been worked out for different populations, and include those for Northern Chinese (Stevenson, 1929), male Germans (Breitinger, 1937), Finns (Telkkä, 1950) and for American Whites and Negroes (Dupertuis & Hadden, 1951). Trotter & Gleser (1952) have computed formulae for American Whites and Negroes, refining the male equations for these two groups in a later publication (1958) and adding tentative equations for the estimation of stature in American mongoloids, Mexican and Puerto Rican males. It is important to realize when estimating stature that the formula that is most appropriate depends upon the group involved. Thus, applying one of the available

Table 5

Regression equations for estimation of maximum, living stature (cm) of American Whites, Negroes and Mongoloids, in order of preference according to standard errors of estimate.

(Data by Trotter & Gleser, 1952, 1958)

Males	Females
White	*White*
1.31 (Fem + Fib) + 63.05	0.68 Hum + 1.17 Fem + 1.15 Tib + 50.12
1.26 (Feb + Tib) + 67.09	1.39 (Fem + Tib) + 53.20
2.60 Fib + 75.50	2.93 Fib + 59.61
2.32 Fem + 65.53	2.90 Tib + 61.53
2.42 Tib + 81.93	1.35 Hum + 1.95 Tib + 52.77
1.82 (Hum + Rad) + 67.97	2.47 Fem + 54.10
1.78 (Hum + Ulna) + 66.98	4.74 Rad + 54.93
2.89 Hum + 78.10	4.27 Ulna + 57.76
3.79 Rad + 79.42	3.36 Hum + 57.97
3.76 Ulna + 75.55	
	Negro
Negro	0.44 Hum − 0.20 Rad + 1.46 Fem + 0.86
1.20 (Fem + Fib) + 67.77	Tib + 56.33
1.15 (Feb + Tib) + 71.75	1.53 Fem + 0.96 Tib + 58.54
2.10 Fem + 72.22	2.28 Fem + 59.76
2.19 Tib + 85.36	1.08 Hum + 1.79 Tib + 62.80
2.34 Fib + 80.07	2.45 Tib + 72.65
1.66 (Hum + Rad) + 73.08	2.49 Fib + 70.90
1.65 (Hum + Ulna) + 70.67	3.08 Hum + 64.67
2.88 Hum + 75.48	3.31 Ulna + 75.38
3.32 Rad + 85.43	3.67 Rad + 71.79
3.20 Ulna + 82.77	
Mongoloid	
1.22 (Fem + Fib) + 70.24	
1.22 (Feb + Tib) + 70.37	
2.40 Fib + 80.56	
2.39 Tib + 81.45	
2.15 Fem + 72.57	
1.68 (Hum + Ulna) + 71.18	
1.67 (Hum + Rad) + 74.83	
2.68 Hum + 83.19	
3.54 Rad + 82.00	
3.48 Ulna + 77.45	

formulae to British skeletal material, a European or American White equation is more likely to yield a reliable stature estimate than, say, one for Mongoloid or Negro groups. However, no physical anthropologist is fully content with the statistics so far produced. As Trotter & Gleser (1952) point out, it is possible that different equations may be needed even for the same racial group in successive generations; moreover, modern formulae may bias the stature estimates of earlier populations.

For the present we have to be satisfied with the formulae available to us. The series of equations that the archaeologist is generally advised to use were devised by Trotter & Gleser (1952, 1958) and are given in Table 5. In the case of early Amerindian populations alternative stature reference data are provided by Genoves (1967). Stature is estimated simply by placing the long-bone length in the appropriate part of the formula and then working it out. For example, if the femur of a 'white' male individual is 45.0 cm, we can place it in the equation for male femora thus: 2.32 × 45.0 + 65.53. When this is worked out, it gives a stature estimate of 169.93 cm (nearly 5 feet 8 inches). It should be noted that although the height given is a precise figure, the individual may in fact have been slightly larger or smaller (this is known as the 'standard error' of the estimate). Moreover this 'standard error' is larger for some bones than for others. Trotter & Gleser (1958) have therefore suggested that if the maximum lengths for a number of different bones from the same individual are available, stature should be calculated from the bone which has the smallest 'standard error', that is, the one least likely to deviate from the actual height. The order of preference for long-bones is given in Table 5. Some workers favour the estimation of stature from as many different long-bones as are available for the individual, the final estimate being the average of the findings for each bone.

The estimation of children's heights from ancient immature skeletal material presents a greater problem. Imrie & Wyburn (1958) remind us that many variables affect the length of a limb-bone during growth, so that at present any height estimates on such materials are mere approximations. This is not to say that recording the information is necessarily valueless, but rather that work in this field is for the future, especially since the amount of early children's skeletons is still small.

12   Body form and the skeleton

In living human populations it is possible to classify individuals into certain categories according to their body form. These varieties of human physique illustrated, for instance, in an atlas by Sheldon (1954), may be correlated to some extent with skeletal robustness and form. It would be quite impractical, however, to try to ascertain body-form to any extent in an individual from an excavated skeleton, although reconstruc-

tion of size and proportions is possible to some extent (Masali, 1972). As Krogman (1938a) records, even when the stature of a person is known, it is still very difficult to guess body-weight. It is therefore of no value generally to try to assess physique from bones, although perhaps excessively slender or robust bones demand a brief comment (especially if the skeleton appears to be pathological).

13   Statistical methods

It is an unfortunate fact that although the application of mathematics to biological problems has helped to establish more precise relationships between series of data and to eliminate personal bias, it has tended to make an appreciation by the layman or the worker in another field even more difficult. The few notes that follow are not intended to imply that the archaeologist is expected to undertake a detailed statistical analysis of osteological finds, but rather to acquaint him with the procedures often employed by specialists. References to introductory books on statistics are given in the section 'Some standard works of reference' (p.179).

What is the general trend in the treatment of osteological data? First of all, in the case of one, two or many specimens, there has been the tendency to apply statistical methods, both to fossil and modern bone material, in order to substantiate visual observations. As Le Gros Clark (1955) pointed out, material comparisons between closely related groups (species, subspecies) are most likely to be reliable, whereas the dimensions of widely differing groups (such as anthropoid apes and man) must be compared with far more caution.

The index — a percental relation of two measurements — was an early attempt at roughly expressing the shape of a bone. This is still of value if there is a need to demonstrate the change in proportion of two dimensions. For example, in Fig. 4.9 we see that maximum head-breadth was closer to the length (thus giving a higher cephalic index) in British Bronze Age than in British Neolithic peoples, although we cannot infer from this whether the brain-case was generally rounder or more square in outline.

In a series of thirty or more skulls, the calculation of mean dimensions will give some idea of the average skull shape; and although this mean form is purely hypothetical, it is of value in the study of different groups. It must be emphasized that the data for both sexes should be kept strictly separate, at least when dealing with measurements. Although the average represents to some extent the whole series of observations, an idea of the range of variation within a group is obtained by the standard deviation. However, as these deviations for osteological data are usually similar in most human groups it is not so important to undertake their computation, unless they are needed in further statistical work.

It is necessary in the comparison of means of two or more groups to determine whether any differences between them are purely due to random sampling of the same population. Depending upon the sizes of the samples in

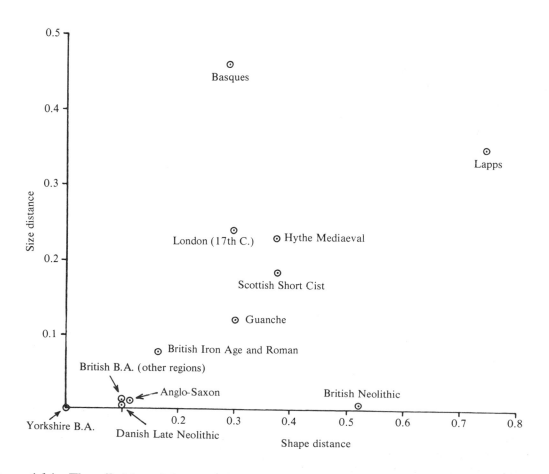

Figure 4.14 The affinities of eleven early and recent peoples with the Bronze Age population of Yorkshire, employing the 'size' and 'shape' method of ten skull measurements. Males only.

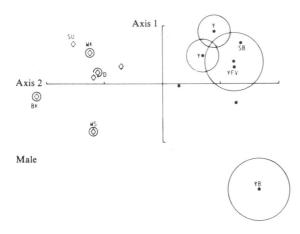

Figure 4.15 Male groups differentiated by canonical axes 1 and 2. Neolithic groups are marked ○, Bronze Age ●. Additional lettering is as follows: SU = southern unchambered Neolithic; WK = West Kennet long barrow sample; D = Dinnington barrow; BK = Belas Knap; WS = Waylands Smithy; Y = Yorkshire samples (without vessels); YFV = bodies with food vessel associations; YB = Yorkshire Beaker burials; SB = southern Bronze Age sample. Encircled ◊ simply indicate the barrow samples.

question, it is usual to solve this problem by calculating the standard error of the difference between the means, or by applying the *t* test of significance.

One of the first attempts to compare more than one or two measurements at the same time was made by Karl Pearson (1926), and this 'coefficient of racial likeness' was applied with some success to various cranial series. The formula has, however, been criticized (Fisher, 1936; Seltzer, 1937). A more efficient method of multifactorial analysis was later devised by Mahalanobis (1936) and Rao (1948). This so-called D^2 statistic was also designed to obtain a measure of divergence between groups in a number of metrical characters. Penrose (1947, 1954) introduced a useful modification of this procedure. This distance statistic may be split into two components or factors which he has called 'size' and 'shape'. The former is a measure of divergence in overall dimensions

and the latter of divergence in the relative magnitude of the measurements. If the populations are being compared by the frequencies of several discontinuous characters, their 'size' measures show divergence in the overall incidence of the characters, while their 'shape' measures show divergence in their relative incidence. Thus the average frequency of the characters might be the same in populations A and B, yielding the same 'size' estimates, but the particular characters which had a high or low incidence might differ between them, thus yielding a divergence in the 'shape' factor. In dealing with continuous metrical characters, it is customary to reduce

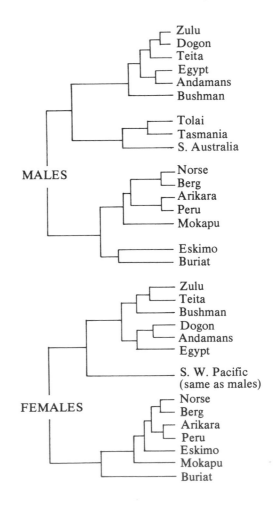

Figure 4.16 Clustering by successive splits.

the set of means to standard measure, obtained by dividing the mean by the standard deviation. In the case of frequency data the raw frequency figures are used.

This latter statistic is of particular value in studying the affinities of one specific group with a number of others. For example, in Figure 4.14 a Bronze Age male series is represented at the origin of the 'size' and 'shape' scales; various other groups are shown at various 'distances' from it

(Brothwell, 1960a). Some results of using an alternative statistical method (canonical analysis) are shown in Fig. 4.15, where again British Bronze Age material is here compared with some Neolithic samples (Brothwell & Krzanowski, 1974).

During the past twenty years, further multivariate analyses have been applied to ancient skeletal material, as exemplified by Howells (1966, 1973), Crichton (1966), Moeschler (1967), van Vark (1970), Henke

Table 6

Measurements used by Howells (1973) in his analysis of human cranial variation. (See also Figs 4.16 and 4.17).

GOL = Glabello-Occipital Length	SOS = Supraorbital Projection
NOL = Nasio-Occipital Length	GLS = Glabella Projection
BNL = Basion-Nasion Length	FOL = Foramen Magnum Length
BBH = Basion-Bregma Height	FRC = Nasion-Bregma Chord
XCB = Maximum Cranial Breadth	FRS = Naison-Bregma Subtense
XFB = Maximum Frontal Breadth	FRF = Nasion-Subtense Fraction
STB = Bistephanic Breadth	PAC = Bregma-Lambda Chord
ZYB = Bizygomatic Breadth	PAS = Bregma-Lambda Subtense
AUB = Biauricular Breadth	PAF = Bregma-Subtense Fraction
WCB = Minimum Cranial Breadth	OCC = Lambda-Opisthion Chord
ASB = Biasterionic Breadth	OCS = Lambda-Opisthion Subtense
BPL = Basion-Prosthion Length	OCF = Lambda-Subtense Fraction
NPH = Nasion-Prosthion Height	VRR = Vertex Radius
NLH = Nasal Height	NAR = Nasion Radius
OBH = Orbit Height Left	SSR = Subspinale Radius
OBB = Orbit Breadth Left	PRR = Prosthion Radius
JUB = Bijugal Breadth	DKR = Dacryon Radius
NLB = Nasal Breadth	ZOR = Zygoorbitale Radius
MAB = Palate Breadth	FMR = Frontomalare Radius
MDH = Mastoid Height	EKR = Ectoconchion Radius
MDB = Mastoid Width	ZMR = Zygomaxillare Radius
ZMB = Bimaxillary Breadth	AVR = M1 Alveolus Radius
SSS = Zygomaxillary Subtense	NAA = Nasion Angle, ba-pr
FMB = Bifrontal Breadth	PRA = Prosthion Angle, na-ba
NAS = Nasio-Frontal Subtense	BAA = Basion Angle, na-pr
EKB = Biorbital Breadth	NBA = Nasion Angle, ba-br
DKS = Dacryon Subtense	BBA = Basion Angle, na-br
DKB = Interorbital Breadth	SSA = Zygomaxillare Angle
NDS = Naso-Dacryal Subtense	NFA = Nasio-Frontal Angle
WNB = Simotic Chord	DKA = Dacryal Angle
SIS = Simotic Subtense	NDA = Naso-Dacryal Angle
IML = Malar Length Inferior	SIA = Simotic Angle
XML = Malar Length Maximum	FRA = Frontal Angle
MLS = Malar Subtense	PAA = Parietal Angle
WMH = Cheek Height	OCA = Occipital Angle

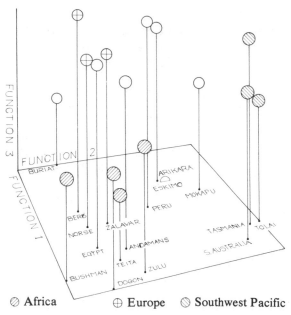

Figure 4.17 A three-dimensional model of population distributions on the first three functions of the multivariate analysis, using the same scale values as Table 6. It shows the separate clusterings of populations in three important regions registered by these three leading functions taken together, also such features as the isolation of the Buriats and Eskimo, and intermediate placing of certain others (after Howells, 1973).

(1972), Andrews & Williams (1973), Constandse-Westermann (1974), Hills & Brothwell (1974), and Brothwell & Krzanowski (1974).

It is not the place here to discuss further the variety of statistical methods that can now be used. A consideration of the literature given will provide further reading for those wishing to go into more detail. The important fact is of course that these multivariate techniques are a means of evaluating the relationships and distances of specimens of populations from one another. The study by Howells (1973) shows the variation occurring just in more recent peoples. His measurement list is especially extensive (Table 6) although some are not in fact in common use. The type of visual patterns reflecting biological distance that can be constructed from the basic craniometric data when analysed, can be shown in the alternative Figures 4.16 and 4.17. The first shows a tree-like clustering of various populations (males and females separately). The second is a three-dimensional model of the same groups, establishing the relationships with more accuracy, but perhaps not visually so easy to understand.

14   The problem of unreliability in samples

It is sometimes very difficult to know whether all the individuals in what appears to be a fairly homogeneous community really belong to the same ethnic group or a section of it. Also, a series of burials from one cemetery may not represent the population as a whole. This problem of samples is of course encountered in studies on modern populations, but is far more difficult in ancient material. There are at least six factors that may bias the series and make it atypical.

i) *Battles*
   a) The males selected for most armies tend to be the fittest, and possibly the most well-built individuals in the community.

b) Professional troops belonging to one army may nevertheless be of various nationalities.

ii) *Plague*
Highly infectious diseases may strike one part of a community to a greater extent than another, either through closeness of living quarters or inability of the individuals in some circumstances to resist disease (e.g. in a crowded slum area).

iii) *Starvation*
As shown by Ivanovsky (1923) and Acheson *et al.* (1956), malnourished individuals may show stunted development, with noticeable reductions in stature.

iv) *Religious isolation*

The burials may represent an interbreeding part of a larger community, segregated for example through religious differences (as with the Jews).

v) *Immigrants*

a) The individuals may be from an area, such as a part of a town, where new immigrants predominate, although they may be in a minority when considering the total population of the region.

b) Contrary to *a*), the group may represent the few surviving indigenous individuals living on in a cultural backwater.

vi) *Slaves*

A cemetery may be purely for a slave community, or for the indigenous group together with their imported slaves.

There is, unfortunately, no easy rule for ascertaining the reliability of a series, and it is therefore important to describe as clearly as possible the circumstances of the burials, especially if it is suspected that the cultural artefacts are atypical of the period and area. If the group may represent one social stratum, or a specific sect, this too should be noted. Battle victims are, of course, usually the easiest to distinguish.

15   'Race' and the skeleton

'An experienced physical anthropologist', wrote Hooton (1947), 'ought to be able to determine from the examination of the skeleton whether it belonged to a "civilized" or an "uncivilized" person, and whether that person was a White, a Negroid, or a Mongoloid. There are also some especially distinctive races, sub-races, or peoples that are usually identifiable from the skeletal characteristics.' This statement is to some extent true, for there are some skeletal features, especially of the skull, that allow us to distinguish quite a high proportion of individuals of one major racial group from those of another (Giles & Elliot 1962, Howells 1973), e.g. the broad flat face of the Eskimo skull cannot easily be confused with that of an Australian aborigine or a negro. In all groups there are skulls that are rather 'neutral' in form and cannot easily be placed in a major category. On the other hand, there are skulls which display, say, the negroid character of strong alveolar prognathism, or the flat, broad cheeks of a mongoloid, and yet do not belong to either group.

Although some skeletal features differentiating the major groups are quite marked, others are more subtle in form and require first-hand experience of various morphological details. Fortunately, the excavator is not often faced with the problem of determining the major racial group of a skeletal series, and for this reason, more detailed descriptions will not be given. Useful references in connection with the identification of skulls are Hooton (1947), and Briggs (1958). Regarding the statistical differentiation of racial samples, Giles & Elliot (1962) have demonstrated by means of discriminant function analyses, that skulls can be separated by measurements into different population clusters.

Classification of skulls into one of the so-called human 'sub-races' is a far more difficult procedure. Such terminology also appears to have only a limited value when dealing with earlier material, and rather than place the specimens in a general category, it is more important to treat the smaller populations separately and to ascertain their distances (metrical and otherwise) from various other relatively small communities both past and contemporary. For example, when considering the origin of the Etruscans, we can, by measurement and statistical analysis, determine their morphological affinities with other groups in the region, without having to consider whether they are all of so-called 'Mediterranean' or 'Nordic' sub-racial type. It would therefore seem advisable, when dealing with excavated skeletal material, to think in terms of the area and archaeological period concerned as a preliminary to more detailed analysis, and not to worry about general racial classification.

If the excavated individuals are thought to be the result of the mixture of different physically contrasting groups, it may be of interest to suggest tentatively which of the two, or more, groups seemed to have had the more influence on skeletal morphology. In the case of a large series, it is worth computing means in order to compare them with the parental populations. Trevor (1953) had showed that hybrid groups often display measurements that fall between the communities from which they were derived.

16   What can be deduced from a single skull

Le Gros Clark (1957) criticized the over-confidence of earlier anthropologists who were often prepared to draw far too many conclusions from a single specimen. Singer (1958), in a critique of the Boskop Skull which was discovered in 1913 on a Transvaal farm, points out that what may have been justifiable speculation at the beginning of the century is inexcusable now. Even the sexing of fossil human remains appears to be far more difficult than was previously imagined, as Genoves (1954) has emphasized in his study of Neanderthal remains. This is not to say that a single find is valueless. Certainly in the case of Palaeolithic remains, each bone demands rigorous examination and description. In later man, where it is necessary to study less outstanding differences between groups, series of at least forty or fifty skeletons of the same sex are needed before reliable statistical analyses are possible. Thus, if only one or two dated skeletons are excavated, sex, age, disease and measurement may be noted but deductions as to the morphological affinities of these individuals with other peoples of the period could only be very tentative, if they are possible at all. The value of reports on one or two skeletons is not to serve as the basis of any theory of morphological relationships, but to provide standardized data that may eventually be built into a general picture of population at one period or through time.

Dating a skull on the morphology has also severe limitations. In the case of British archaeological remains, if there is some doubt whether a robust brachycephalic skull is of Neolithic or Bronze Age date, one may say that there is a greater likelihood of it being Bronze Age, for most Neolithic individuals were noticeably long-headed, whereas many Bronze Age invaders were very short-headed. However, we cannot be sure, for it is clear that from Mesolithic times onwards the frequency of brachycephaly was increasing continually in some parts of Europe. After Bronze Age times, British groups became more heterogeneous with each successive 'invasion', and although some skull 'types' or 'forms' often predominate in a series, morphological distinctions between populations must rely even more on statistical analyses of measurements.

Can the major race affinities of an individual easily be determined? The answer is that an isolated skull cannot always be so classified. Even if it is possible, the individual may be assigned only to the broadest concept of 'racial stock'. Although evidence of 'race mixture' in earlier times is of considerable interest − it is a field that has been quite neglected, and is deserving of far more study.

17 The question of family relationships

It is sometimes of particular interest to the archaeologist, especially when dealing with material from burial chambers or the remains of small isolated communities, to know whether the skeletons represent a fairly closely related group, such as a family unit or small interbreeding community. Unfortunately, the identification of such groups is far from easy, owing to the fact that so little is known about the inheritance of commonly occurring skeletal forms. Thus, although a number of skulls may look very similar, the limited information available does not allow us to judge on these grounds whether the individuals were related. Similarly, the occurrence of common anomalies, such as wormian bones, is also slender evidence. It is to the more unusual anomalies of a congenital nature that we must look for more reliable evidence, and even some of these must be regarded with reserve. A few examples from earlier times will perhaps serve to show what sort of anomalies may make a claim to 'family likeness' feasible.

The first case is seen in British archaeological material. Since 1937, twelve Romano-British skeletons have been uncovered at Arbury Road, Cambridge. Surprisingly, five of them showed sacralization of the fifth lumbar vertebra (Fig. 4.18), principally on the right side (Fell, 1956). Although the high frequency of this anomaly cannot be used as absolute proof of a family relationship, it does suggest close interbreeding. Another collection of five similar anomalies, this time from Nubia, also suggests a close genetic relationship. Elliot Smith & Wood Jones (1910) described five females of the Christian period, each with a noticeable anomaly at the hip (concerned with a congenital abnormality of the femoral head). It is remarkable that all were discovered in one small district, three in the same cemetery, and two actually in adjacent graves. In the British Museum (Natural History), a small series of Saxon burials from Guildown show a number of defects of the sacrum; probably five specimens (three for certain) display an open sacral canal on the dorsal surface (Fig. 4.18). Post (1966) has considered this sacral condition, *spina bifida occulta* in various population samples, and has concluded that higher frequencies of the defect occur in more 'civilized' communities.

Another example, this time from a more recent period, is particularly well-authenticated. The jaw deformity known as Angle's Class III, consisting of an excessively long lower jaw and a poorly developed upper one, is often found in families, and the mode of inheritance is apparently dominant. This anomaly was manifest in the House of Hapsburg from A.D. 1377 until after 1700 (as can be clearly seen in portraits).

Perhaps the most interesting case is that related by Stern (1950). The anomaly known as brachydactyly or short-fingeredness, in which the bones (phalanges) of the fingers are reduced and some may even be fused or missing, is known to be inherited as a simple dominant. In both the first Earl of Shrewsbury (born 1390), whose skeleton became available for study, and in a nineteenth-century descendant of his line, fusions of certain phalangeal bones were noted.

Abnormalities of this nature do not occur very frequently in skeletal material; so that until more is known about the inheritance of commonly occurring features, deductions as to family relationships within groups must in general remain rather tentative.

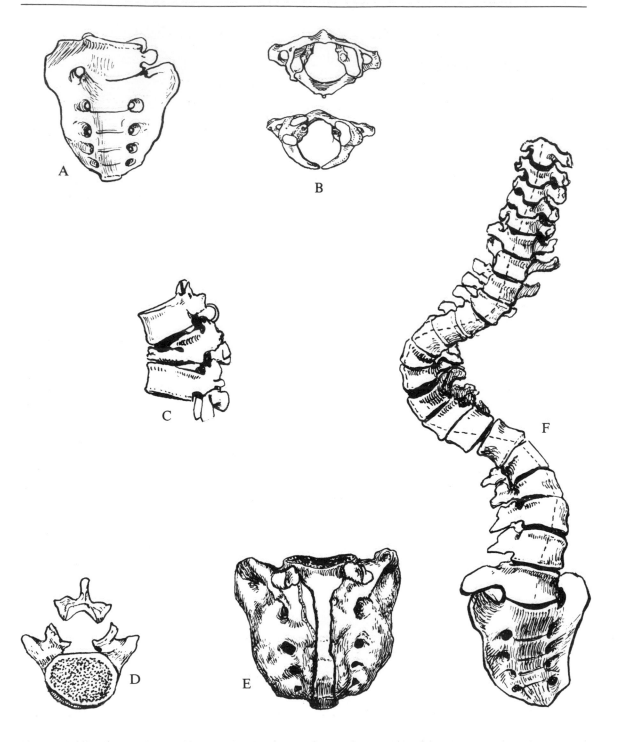

Figure 4.18   Anomalies of the vertebral column. A, partly sacralized lumbar vertebra. B, normal and anomalous atlas vertebrae from Lapps. C, collapsed vertebra of an Eskimo (after T. D. Stewart). D, separate neural arch in a lumbar vertebra of a Lapp. E, congenital exposure of the sacral canal in an Anglo-Saxon. F, well-marked vertebral scoliosis (after J. P. Weinmann & H. Sicher).

18   Teeth and archaeology

The study of the teeth in earlier man is just as important as the metrical and morphological study of the skull bones. Since teeth are particularly resistant to decay when buried, they survive much longer in some soils than bones.

It is clear that teeth vary in many ways. The individual experiences the eruption of two different sets in the course of his life. There are intra-group differences to be seen in each set. The teeth vary considerably between man and other primates, between fossil and recent man, and between relatively modern groups.

The sequence of dental eruption facilitates age-estimation of the young in burials, while age-changes in the permanent adult teeth permit further estimates. Teeth are therefore of considerable value in estimating mean ages of early populations. Tooth size may sometimes help in sexing fragmentary specimens, which in turn helps to build up a picture of the frequency of the sexes in certain types of burial monument or cemetery. The mean shapes and dimensions of teeth may vary between groups and show evolutionary trends or ethnic affinities. Congenital variations in the number of teeth (especially third molars) may suggest slight genetic differences between large groups, or could perhaps point to an isolated village or tomb community being closely interbreeding.

A consideration of the numbers of teeth, taking into account age factors (degree of closure of the pulp cavity, or of attrition), shape and size differences, is often of value in determining the number of individuals present in an excavation, especially if the remains are poor and scattered. Humphreys (1951), for example, discusses the value of teeth in sorting the human remains from the Bredon Hill Iron Age fort. It would appear that those slain in battle were never buried, with the result that there was a complete scattering of the bones. However, by counting the first left lower molars, it was shown that no less than fifty-four men had died there, and this estimate agreed with the more laborious comparison of bones.

Differences in the incidence of dental decay may suggest diet variations in early populations, even where there is an absence of cultural evidence in this respect. Inferior enamel, especially if there is a high frequency of hypoplasia within the group, might be correlated with periods of malnutrition or disease during childhood. Unlike bone, where faulty calcification for a period may be completely eradicated by the process of repair, teeth always preserve evidence of an early phase of ill-health provided the dental calcification was not complete at the time. Such evidence of inferior calcification is certainly not uncommon in archaeological material and the excavator should make as much use of dental evidence as possible.

19   Variations in tooth number

Even after accounting for some difference in numbers through loss of teeth as a result of decay during life, anomalies may be noted that are developmental in origin. Such differences from the normal may be controlled by genetic factors or, as certain work on rodents suggests, 'environment' may sometimes have played its part.

i) Absence of teeth

Lack of the third molar is the commonest example of missing teeth to be found in archaeological material. Ethnic variation in the frequency of absence of this tooth has not been extensively studied, but there are evidently differences. A deficiency of third

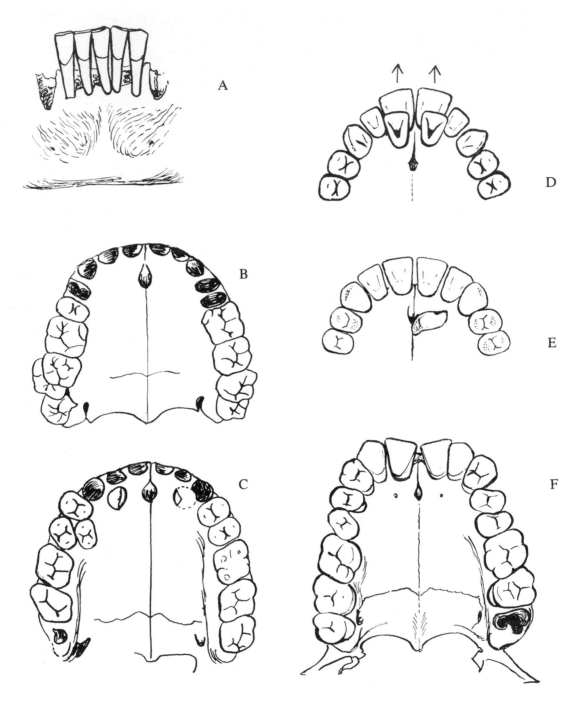

Figure 4.19 Variations in extra (supernumerary) teeth. A, presence of five incisors in a mandible from Thames deposits. B, bilateral occurrence of fourth molars in an Ashanti. C, Nigerian female specimen with an extra premolar and bilateral supernumeraries buried within the palate. D, additional medial incisors in a Prona Indian. E, pre-medieval British case of a non-specific medial supernumerary tooth on the surface of the palate. F, presence of extra premolars and absence of canines in a Nepalese specimen.

molars has been found to be as high as 20 per cent in civilized groups (Banks, 1934). Distinction must be made between teeth that have developed but are unerupted, and those that have not formed at all. A final decision on this point may demand radiographic assistance, but it is usually evident from the positioning of the other teeth, and the condition of the alveolus, whether a tooth is congenitally absent. If a permanent tooth has been lost after eruption, the space is generally larger than if the tooth is undeveloped, and the alveolus tends to be more unequal and resorbed in this area. Moreover, if the loss has taken place after the eruption of the adjoining teeth, it may be found that they have moved in at an angle and may partially fill the gap. In the case of the third molars, if the first and second molars have been lost through disease, it may be very difficult to be sure whether the former were ever present or not.

All the other types of teeth can be missing congenitally. Oliver, Brekhus & Montelius (1945) note that the absence of second premolars is also fairly common, at least in modern material. Absence of incisors has been found to be much more common in Mongoliform groups, including Eskimos (Pedersen, 1949), than among Europeans. 'Partial' anodontia, meaning the development of only a few teeth, or 'total' anodontia where no teeth are formed, are rare phenomena.

Information about the frequencies of congenitally missing teeth in earlier populations is still sadly incomplete, although a few frequencies are available (Brothwell, Carbonell & Goose, 1963).

ii) Extra (supernumerary) teeth

The reverse to the previous condition is the presence of more teeth than usual: one or more extra incisors, canines, premolars or molars (Fig. 4.19). Sometimes the increased number is caused by the retention of a milk tooth into adulthood (Fig. 4.20), even though all the permanent teeth were able to erupt. The incidence of extra teeth is usually considered to be about one-tenth of that of missing teeth. They may erupt normally and resemble in every way an adjacent normal tooth; on the other hand they may be improperly placed and mis-shapen. If there are two similar extra teeth, it is common to find them symmetrically placed at each side of the jaw (even if out of line with the other teeth). More rarely the deciduous dentition may contain extra teeth.

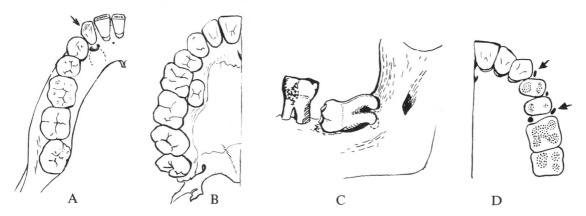

Figure 4.20   A, retention of a milk canine into adult life, with the development but retention of the permanent canine within the jaw. Bronze Age specimen from Dorset. B, retention of a milk molar and eruption of the permanent 2nd premolar to one side of it. Andamanese. C, early British case of horizontally positioned third molar. D, retention of the roots of milk teeth by the side of permanent teeth in a Guanche.

20 Variation in tooth position

In general, the teeth of pre-medieval people in Europe, as well as in modern primitives, show far less evidence of irregular positioning than the teeth of modern civilized populations. However, as Brash (1956) insists, it must not be thought that irregularity did *not* exist in ancient populations. Indeed the evidence suggests that all the abnormalities of this kind known today, occurred to some extent in Neolithic and probably earlier times. For example, cases of incisor crowding, tooth rotation and impacted third molars are shown by Ruffer (1920) in Ancient Egyptian jaws, and similar cases have been found in British Neolithic material. Other types of irregularity known to have occurred in earlier times are shown in Fig. 4.21. Unfortunately, few detailed studies have been made of ancient populations and much work remains to be done before it will be possible to trace the development of our own, often malformed jaws. The only study of British material specifically concerned with early types of occlusion is by Smyth (1933), who found that in a group of sixth-century Saxons from Bidford-on-Avon, 53 per cent showed obvious malocclusion. All the usual varieties were present with the exception of open bite.

While considering the position of teeth we should remember the so-called edge-to-edge occlusion or bite seen in the incisor region. This feature is seen in many skulls of Saxon and earlier times, but by the seventeenth century there was a change to a greater frequency of over-bite. This change probably depended upon other minor alterations in jaw growth and size, and reflects the general decrease in attrition as we come into modern times. The latter factor may well be the more important, for over-bite predominates even in the children of groups where there are many cases of edge-to-edge bite in the adults. In other words, as the teeth wear down, there is a tendency for the occlusion to change slightly, and this is especially evident in the anterior part of the jaw.

Although abnormal positioning of one or more teeth may be easily noticed and described, the difficulties in defining and analysing finer grades of malocclusion make it necessary to leave this task to the dental specialist. Some minor degrees of malocclusion, as for instance the bilateral rotation of maxillary central incisors, seem to have a genetic basis and show population differences (Escobar, Melnick & Conneally, 1976).

21 Abnormalities of tooth form and size

The morphology and size of a tooth are governed by genetic, developmental and post-eruptive factors. Of these the most important is the influence of heredity, although the number of genes involved in forming the dentition is not known. A number of morphological features that differ within and between populations may be briefly discussed as probable examples of genetic traits:

a) Taurodontism

This anomaly, which is found in molar teeth, results from the 'body' of the tooth enlarging at the expense of the roots (Fig. 4.21). Keith (1913) discussed it in detail, considering it to be a characteristic feature of Neanderthal man. Middleton Shaw (1928, 1931) discussed taurodontism in living South African peoples and divided molar root-form into four types: i) cynodont (normal), ii) hypotaurodont (slight), iii) mesotaurodont, iv) hyper-taurodont. The more extreme forms of taurodontism appear more likely to occur in fossil man. As yet few modern cases have been found in Europe, but examples have been described by Miles (1954) and Lunt (1954). It is known to occur more commonly in South African and Mongoloid people (Middleton Shaw, 1928; Tratman, 1950).

b) Shovel-shaped teeth

Sometimes the lingual (inner) surface of the crown of the maxillary incisors has a concavity with a central fossa. Because of the general resemblance of this crown-form to a shovel, Hrdlička (1920) coined the term 'shovel-shaped incisor' (Fig. 4.21). Such

incisors have been found both in modern and early populations, though data on frequency are as yet mainly restricted to fairly recent groups. High frequencies (even over 80 per cent) have been found in Mongoloid groups by various authors, the condition being particularly prevalent in the median incisors. The incidence is far lower in Europe, where moderate to pronounced degrees of this anomaly may occur in about 15 per cent of individuals. Little has yet been done in

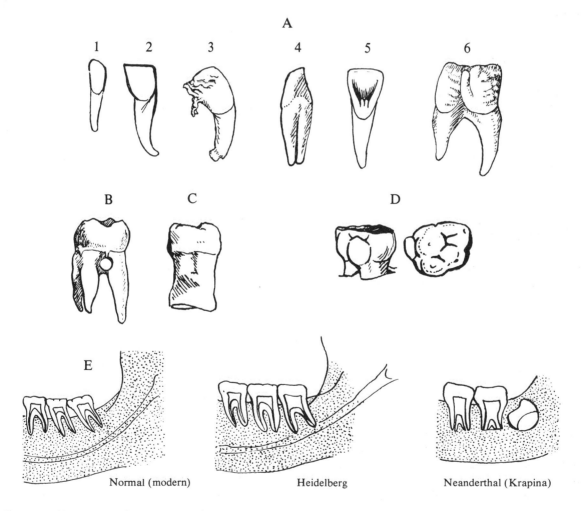

Figure 4.21   A, variations in the form of the anterior teeth. 1, very reduced incisor; 2, incisor with very curved root; 3, incisor with malformed crown; 4, canine with two roots; 5, an incisor with a markedly shovel-shaped crown; 6, fused anterior teeth. B, molar with enamel 'pearl' at roots. C, an extreme case of taurodontism. D, well-defined extra cusp on molar. E, radiographic appearance of normal and taurodont teeth.

studying the occurrence of shovel-shaped incisors amongst archaeological remains, although Carbonell (1963) does include some early material in her study.

c) Number and pattern of cusps

Cusp patterns are of interest to the anthropologist, not only on account of variation in modern human groups, but because they show marked changes throughout primate evolution. In Recent man variations in cusp number are particularly open to analysis. A useful review of work on this subject has been given by Lasker (1950). Little evidence is yet forthcoming on the variability of these characters in early European material, either as regards the normal four or five cusps or concerning supernumerary ones such as the cusp of Carabelli and the paramolar tubercle (Dahlberg, 1945).

d) Other variables

Less important differences in tooth form produced during development may briefly be mentioned. Enamel margins show varying degrees of extension down towards the root (Moeschler, 1968), and areas of enamel are occasionally present as separate nodules or 'pearls' (Fig. 4.21).

Extra roots may be present not only in molars, but also in the anterior teeth, and Duckworth (1915) claims that double-rooted canine teeth are peculiarly frequent in early British crania.

Environment also plays its part in determining tooth form and size. The wisdom teeth (third molars) in both early and modern man are often noticeably smaller than the other molars. Experiments on certain mammals suggest that this may be dictated to some extent by environmental conditions, though whether intra-uterine or at a post-natal time is debatable. Trauma also affects the developing tooth, producing distorted shapes (Fig. 4.21). Disease may produce hypoplastic forms (discussed on p. 159), and syphilis may give rise to characteristic deformity (Fig. 4.22).

Excluding disease and attrition, any changes in shape during the post-eruptive period are usually attributable to cultural factors. Occasionally grooves are made, or areas of tooth removed by accident. Most cases, however, are the result of intentional deformation; either the result of filing or of chipping the teeth. Numerous surveys of dental mutilation in different groups have been undertaken, contributions including those of Singer (1953a) and Comas (1957). Cases are known from Africa, America, Malaya, Australia and even Ancient Egypt. A wide variety of mutilations are known, of which a selection is illustrated in Fig. 4.22A. In some early cultures, and especially in Mesolithic-Neolithic North Africans, mutilations include tooth evulsion (removal in life) for ritual purposes. Examples of this practice (Fig. 4.22C) have been noted in skulls from Afalou bou Rhummel, Assalar, Taforalt, Mechta el Arbi and other sites (Verger-Pratoucy, 1968).

A   Variations in the form of artificially mutilated teeth. From various parts of the world.

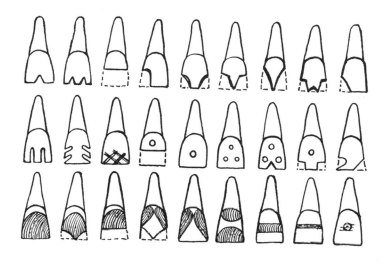

A normal
variant

Deformity
associated
with
congenital
syphilis

Accidentally
chipped (ante-
or post-mortem)

Accidental
wear
caused
by a clay pipe

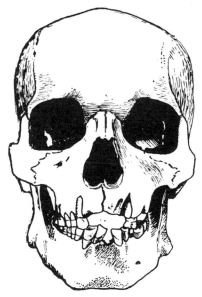

B Anomalous tooth forms
   which could be mistaken
   for cultural deformation

C A typical case of tooth evulsion in a
   Mesolithic man from North Africa
   (Mechta 8). After Cabot Briggs, 1955.

Figure 4.22   Variation in incisor teeth resulting from disease, accident or mutilation.

V Injuries and marks on bone

1 Interpretation of bone injuries (Trauma)

Ever since man evolved he appears to have asserted himself by employing violence and this is particularly evident from the discovery of various bone injuries in archaeological material. Courville (1950) reviewed the evidence for cranial injury in prehistoric man, noting its presence in the very early *Homo erectus* group as well as in the later Neanderthal and Cro-Magnon forms of man. Presumably, as there would have been relatively few accidents in early times, especially before large cities developed, most of the injuries probably resulted from intentional blows. However, falls must have led to fracturing at times, and Cameron (1934), for example, found a number of broken clavicles in Neolithic and later prehistoric material. A preliminary classification of weapons (modified from Harrison, 1929) and the types of injury they cause can be given as follows:

a) Gross crushing by large stones or clubs.
This may result in considerable deformation, and in the case of the skull, may show a primary depressed area from which subsidiary cracks radiate. It may be noted here that baboon skulls found at some Australopithecine sites display compressed fractures that were probably caused by bone or stone implements.

b) Less extensive fracturing by small clubs, maces and missile-stones.
The individual is far more likely to have survived with these smaller injuries, with the result that signs of healing may be detected. In the case of the skull, broken noses are fairly common (Fig. 5.1). Various degrees of long-bone fracturing may be noted. Although these often resulted in deformity in earlier man, a few of the breaks appear to have been well treated, and indeed, splints were used even in ancient Egypt (Elliot Smith, 1908).

c) Piercing by spears, lances, daggers, javelins and arrows.
Definite cases of perforation may usually be distinguished from post-mortem erosion by the well-defined shape of the hole (unfortunately, the excavator's pick or trowel may sometimes produce very similar effects, although the differences in bone colour usually make such cases obvious). Sometimes, as in the Iron-Age Skeleton P7A at Maiden Castle (Wheeler, 1943) an arrowhead actually lies buried in the bone. In the case of the skull vault, the perforation may be larger on the endocranial surface than externally, a feature that has been noted in the circular opening through the left temporal squama of the Rhodesian skull.

d) Cutting by swords and axes.
The wounds may be narrow, if resulting from the sword, or wider if a blunter instrument was used. With the skull, there may or may not be associated secondary cracks (Fig. 5.1). The injuries may take the form of long deep gashes, shorter and more superficial incisions, or minor scratches caused by the sword glancing off the bone. Sometimes roundels of bone are removed from the skull by cuts (see p.122). As seen in Iron Age specimens from Sutton Walls (Cornwall, 1954), decapitation usually results in cleanly divided cervical vertebrae.

As well as these various forms of ante-mortem injury, post-mortem damage must be considered. Sometimes it is very difficult, if there is no sign of healing along the fracture lines, to be sure that the injury is not purely the result of post-mortem crushing through earth-pressure. Occasionally the bones may exhibit scars made by the teeth of rodents. Bennington (1912), for example, in a series of Negro skulls, noted damage due to gnawing (Fig. 5.2), and similar markings are to be seen

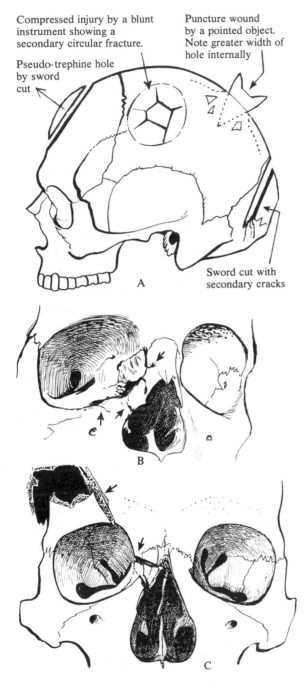

Figure 5.1   A, Variations in the type of cranial injury. B, Injury to the orbital and nasal regions, showing considerable healing. Gaboon male [1864.67.36]. C, Injury to the face received just before death, showing the clear-cut unhealed nature of the wounds. Nigerian female [AF 1822].

on human bones from the island of Socotra and in British Neolithic material.

Two other forms of marking need only brief mention here. Firstly, the trephining of skulls is undertaken both on the living and on the dead, and will be discussed more fully in the following section. Secondly, bones may be found that have been mutilated for cultural reasons. Wankel (1883), for example, mentions the early use of the human brain-box as a cup (Fig. 5.3); and instances of this practice in modern times are known. Lumholtz & Hrdlička (1898), found on the long-bones of prehistoric Tarasco Indians, various parallel markings and artificial perforations, presumably originating as orna-mentation. Decorated skulls have also been found including a particularly artistic example from Guatemala (de Terra, 1957), which is illustrated in Fig. 5.2.

2   Trephining or trepanning

Probably one of the earliest operations performed by man was on the vault of the skull. This consisted in making a number of incisions into one or more cranial bones and usually resulted in the removal of a rectangle or disc of bone (sometimes described as a roundel). This operation is generally called trephining or trepanning, the aperture result-ing being known as a trephine hole. Occasionally this is confused with a con-genital condition known as cranial dysra-phism (Stewart, 1975).

The first case of prehistoric trephining to be recognized was in a skull from Cuzco, Peru, noted by the anthropologist E. G. Squier while visiting Peru in 1863–65. Brocka's interest in this find led to the description of trephined French Neolithic skulls. Since then many such cases have been described and discussed, more important contributions including the survey of European cases by Piggott (1940), the general reviews by Stewart (1958a) and Lisowski (1967), and further prehistoric European examples considered by Ullrich &

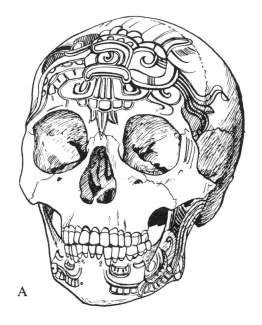
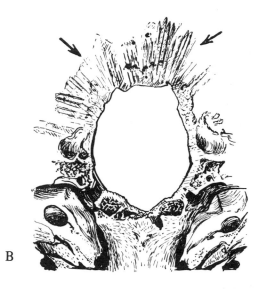

A   B

Figure 5.2   Decorated skulls, Guatemala

Weickmann (1965) and Nemeskéri (1976*b*).

Examples of trephining are known from Europe, the Pacific, South America, North America, Africa and Asia. The operation is also known to have occurred in fairly recent times in North Africa, South America, Serbia, Melanesia and Polynesia. In the British Isles there are at least twelve probable cases of trephining, each skull showing but one hole. The relevant sites of discovery are shown in Fig. 5.4. Various motives for undertaking this dangerous operation have been suggested, as follows:

a) For the purpose of obtaining roundels for use as amulets, either after death or from living captives.

b) As a surgical treatment of an injury especially in the case of fractures of the skull.

c) As a medicinal procedure to combat headaches, epilepsy and other illness.

d) Crump (1901) noted that in New Ireland the operation had become fashionable with no reason other than that it was considered to be an aid to longevity.

A   B

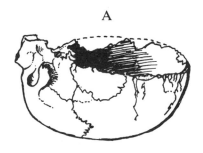
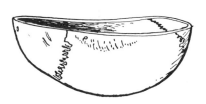

Figure 5.3   A, A prehistoric example of a skull used as a container. B, A modern example, probably from Africa.

As Stewart (1958*a*) says, it is very likely that the most frequent reason for trephining in Peru and Melanesia was for the alleviation of pressure on the brain caused by skull fracture. Other reasons accounted for fewer cases in these areas, although probably for more in other regions. The practice of obtaining roundels of human skull bone as amulets, presumably for magico-religious reasons, seems to have been undertaken in prehistoric Europe, and is practised in parts of Africa today. Very occasionally the amulet may be found still in position, as exemplified by the Crichel Down skull from Dorset (Piggott, 1940). Sometimes the pieces of skull were polished or perforated for suspension.

If a skull appears to show a trephine hole, a careful examination should be undertaken. The shape and size of the perforation, evidence of tool marks (often seen as deep scratches), and any signs of healing should be noted. Healing may be considered to have

taken place if the edges of the hole are rounded and the exposed diploic spaces of the spongy inner table show signs of closing or have closed. Not all such perforations are necessarily man-made, and the following conditions may also produce holes or marks very similar to those resulting from trephining:

a) The removal of a 'roundel' by a sword cut, as in the skulls from the Stanwick Fortifications (Pl. 5.1) (Osman Hill, 1954) and Sutton Walls (Cornwall, 1954).

b) Holes made by a pick during excavation. This may be fairly well-defined especially if the bones were wet and soft. Similar injuries may be made by grave-plunderers or in disturbance by later grave-diggers.

c) The continual pressure of a sharp stone. This may be the cause of the hole in the right parietal of Aveline's Hole skull O.

d) Selective erosion of one region of the skull, especially if an area of bone is already broken or crushed.

e) In some parts of the world, the action of beetles or porcupines or other rodents can produce extensive destruction of bone.

f) Occasionally congenital deficiencies of the parietal bones result in thinning or complete circular perforation. Such a case was found in a graveyard at Eastry, Kent (Parry, 1928).

g) Syphilis may sometimes produce an opening looking very similar to an eroded trephine hole (Pl. 5.2).

Three main kinds of trephining technique should be noted (Fig. 5.5), the first being by far the commonest throughout the world.

i) The instrumental scraping of an annular groove, which on penetration of the inner table results in the separation of a roundel. Usually the external circumference of the hole is larger than the internal, to an extent depending on whether the cutting gradient is steep or gradual.

ii) The production of grooves through

Figure 5.4 Distribution of archaeological trephine cases in Great Britain and Ireland.

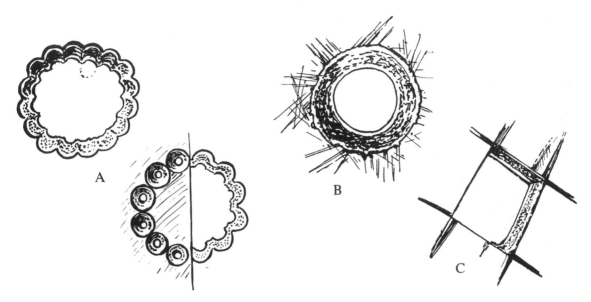

Figure 5.5 Types of trephine hole. A, By drilling holes and cutting through the divisions between them. B, By slowly cutting out a roundel with a metal or flint tool. C, By four deep incisions which enclose a rectangular area of bone.

depressions forming a rectangular pattern. This form is mainly found in Peru but there are one French Neolithic and two ancient Palestinian examples.

iii) By drilling small holes in a circular pattern and then cutting the narrow connecting bars between them. Only Peruvian examples of this type are known.

It has been found that the left side of the head is the commoner side of operation, this bias probably resulting from the fact that injuries to the head through single combat are more likely to be on this side. Moreover, the frontal area is most favoured for the operation, secondly the parietals, and more occasionally the occipital bone. Over 50 per cent of the cases known show complete healing, which is extremely remarkable considering the likelihood of infection, danger of fatal haemorrhage, the crude nature of the operation technique and the danger of severe shock. The number of holes per skull is not necessarily restricted to one, and in a skull from the Cuzco region, now in the British Museum (Natural History), as many as seven healed trephine holes may be seen (Oakley *et al.*, 1959).

Sometimes a circular area around the trephine hole appears to be etched or pitted. This halo of osteitis shows that the individual had lived after the operation and that some degree of infection had set in afterwards. It has been suggested that this is in fact a chemical osteitis, resulting from the applications of medicants to the wound, but I am inclined to agree with Stewart (1956*c*) that it is more likely to be septic osteitis. A good example of this condition is seen in a skull from Lachish (Pl. 5.1).

It may be noted that not all early head operations were in the form of trephine holes and there is some evidence that the primitive surgeons in certain areas produced extensive scars on the skull vault. A number of these have been claimed to be in the form of a cross (Sincipital T).

A peculiar mutilation at the foramen magnum has been reviewed in detail by Kiszely (1970), who considers the enlargement of this foramen to be for medical or cultural reasons. From the published evidence, damage by rodents still seems to be an alternative explanation, and further study is clearly demanded.

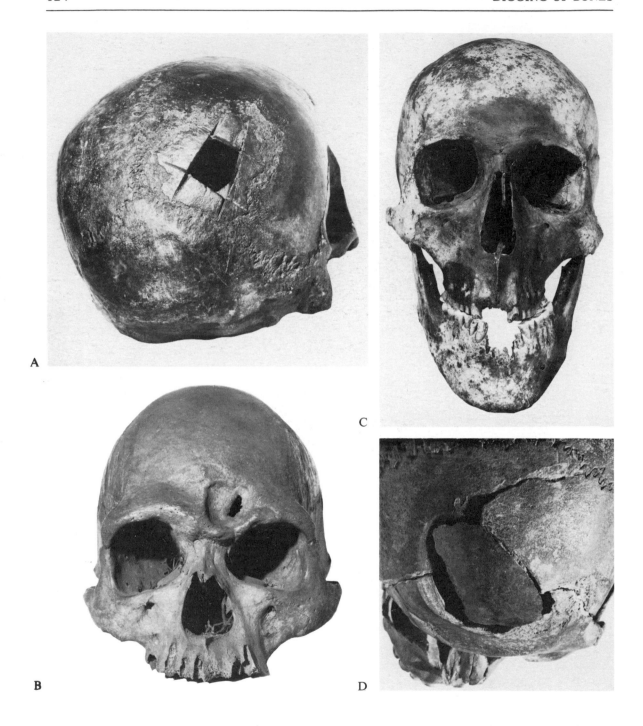

Plate 5.1   A, Posterior aspect of a skull from Lachish, Palestine, displaying the rectangular type of trephine hole. B, Skull from New Ireland with healed frontal trephine. C, Facial view of a female skull believed to be of an Ancient Egyptian, showing considerable facial lenthening, possibly caused by acromegaly. D, The frontal region of a skull from the Stanwick Fortifications with a sword injury simulating trephination.

A   B

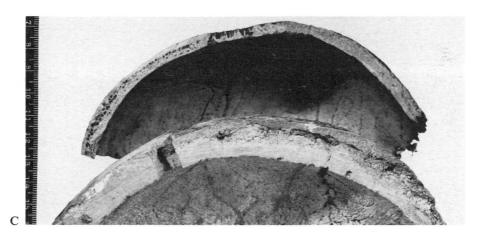

C

Plate 5.2   A, Skull from Spitalfields, London (probably late medieval). Extensive destruction of bone has occurred in the region of the face and upper part of the skull due to syphilis. In the frontal region there is a perforation which simulates a trephine hole. (By courtesy of the Duckworth Laboratory, Cambridge.) B, The left ear and mastoid region of an Early Dynastic Egyptian skull from Tarkhan, showing a perforation of the posterior aspects of the ear hole into the mastoid sinuses, caused by an infection in this region. (By courtesy of the Duckworth Laboratory, Cambridge.) C, Sections of the frontal and parietal bones in a normal individual (upper specimen) and a person who suffered from Paget's disease (both recent specimens).

3   Tooth evulsion

The practice of intentionally knocking out teeth, which may be associated with beautification, an initiation or other age ritual or even a sign of grief, is particularly well known in Africa. Only the front teeth are involved, and usually evulsion is restricted to the incisors. One or both jaws may be deformed in this way (Fig. 4.22). Singer (1953a) has reviewed its presence in various recent African groups, and Briggs (1955) notes its occurrence in earlier material. It would appear that almost the entire Mesolithic population of northwest Africa removed one or more of the incisors, and sometimes the canines, of their young. In isolated cases of missing teeth, we cannot be sure that it is intentional evulsion, for it may be simply the result of disease or a blow to the face; but two or more similar cases in a group point to this practice. Keith (1931), describing remains of the late cave dwellers of Shukbah in Palestine, reported that four out of five reasonably complete palates showed undoubted evulsion.

4   Blood-stained bones

It is extremely difficult to differentiate between early post-mortem injury of bone and some forms of injury received immediately before death. This problem led Elliot Smith & Wood Jones (1910) to look for other features that might enable these ante- and post-mortem conditions to be distinguished. After examining numerous skeletons of early Nubians which showed definite signs of injury at or before death, they formulated the opinion that blood-staining of bone in the vicinity of an ante-mortem injury was in fact retained as an area of deeper staining. They write:

'This blood-staining of the bones is a diagnostic point of great value in bodies buried in Nubia. The limit of its persistence we do not know, but it is certainly very vivid upon bones the archaeological dating of which refers them to a period well over five thousand years ago'. Supporting evidence included colour changes in the mummy of Rameses V, where a 'wide area of discoloration' attributed to blood-staining was noted around an ante-mortem injury to the skull. However, chemical tests applied to such stained areas failed to reveal blood substances.

More recently, Waterston (1927) has discussed similar stains on a male skeleton from a stone cist near Fife. Archaeological dating of this find was not certain, although no doubt it is prehistoric. Here again, the staining on the bone tissue failed to show any definite similarities to blood pigment. However, as Waterston says: 'A little consideration will show that this is what might be expected, and forms no argument against the view that the stain is due to blood, ... in specimens such as those described here, the organic matter has been destroyed, and only the inorganic residue remains.' His argument continued that as iron forms a large proportion of the inorganic residue of blood, then a greater degree of ferruginous staining in the region of an injury is probably due to blood-staining.

Quite clearly this topic demands further analysis, especially now that more refined biochemical methods are at hand. In the case of skeletal (as opposed to mummified) material, a very serious complicating factor is the fact that some soils and other deposits produce patches of darker staining of bone through heterogeneous concentration of ferruginous material.

VI Ancient disease

1 Introduction

It is very important to note in archaeological material any anomalies that have resulted from disease.

Although bone diseases such as osteomyelitis, tumours, dental decay and abscesses have been found in fossil material millions of years older than man, it seems probable that many of the more lethal human parasite strains came into being during the course of man's biological and cultural evolution. Some of these microscopic organisms common now to man and other animals have probably always been associated with them in the past, and Hare (1954), points out that general wound infections, peritonitis resulting from intestinal wounds, as well as such spore infections as tetanus and gas gangrene would probably have been as current then as now.

However, in Palaeolithic and Mesolithic times, at least, the size of the interbreeding social units possibly severely hampered the survival of some diseases. Any highly lethal strain that evolved may be conceived as quickly burning itself out in a small group, further dispersion being generally prevented by the relative isolation of the group.

It is also to be noted that the environment need not necessarily have caused the spread of infection, although malnutrition has evidently been important at times by lowering the body's resistance to infection. A few decades ago, the isolated inhabitants on the island of Tristan de Cunha were particularly healthy even though the climate and sanitation were poor, the diet rough and relatively deficient in vitamins, and the drinking water polluted.

Many diseases do not, of course, involve changes in the skeleton, but those that do are of great value in assessing the health of earlier peoples. From Neolithic times onwards the number of parasites associated with man has probably increased century by century. Whether these early organisms were identical with, and produced similar reactions to the modern forms, are questions whose solutions rest in the study of human archaeological remains.

During the past seventeen years, a growing number of publications have appeared that are concerned with the study of ancient human disease (Wells, 1964; Jarcho, 1966; Brothwell & Sandison, 1967; Morse, 1969; Jansens, 1970; Steinbock, 1976). These vary in detail and approach, but indicate increasing interest and a gradual change to an epidemiological or ecological approach to disease in archaeology.

Although the frequency of some of these diseases is fairly low, the examination of skeletons for signs of anomaly must not be overlooked. Moreover, it is important that sufficient space should be given to this aspect in reports. Palaeopathology and palaeo-epidemiology not only enable various modern ailments to be traced back into history and pre-history, both individually and at a population level, but may also provide new information concerning the manner in which they can manifest themselves.

2   Classification of bone diseases

It is evident that although the rare bone diseases are worthy of description if detected in skeletal remains, a full classification as found in general medical textbooks would tend to be cumbersome here, and might only lead to confusion. As a working division of bone disease, the following is therefore suggested; it is mainly a modification of that given by Fairbank (1951) and Brockman (1948).

i) *General infections of bone*
   Osteitis and periostitis
   Osteomyelitis
   Tuberculous disease of bone
   Treponemal diseases of bone
   Leprosy

ii) *Tumours of bone* (Neoplasms)

iii) *Diseases of joints* (Arthropathies)
   Arthritis ('rheumatism'): *a*) rheumatoid arthritis; *b*) osteoarthrosis; *c*) others

iv) *Diseases of the jaws and teeth*
   Caries
   Periodontal disease
   Abscesses (chronic)
   Hypoplasia
   Cysts
   Odontomes

v) *Deformities*
   Infantile paralysis (poliomyelitis)
   Hip deformities. Congenital dysplasia of the hip

vi) *Bone changes due to endocrine disturbances*
   Hyperpituitarism: *a*) gigantism; *b*) acromegaly
   Hypopituitarism: dwarfism

vii) *Effect of diet on bone*
   Rickets
   Osteomalacia
   Other diseases produced by diet

viii) *Bone changes associated with blood disorders*

ix) *Acquired affections of uncertain origin*
   Paget's disease
   Osteoporosis

x) *Congenital development errors*
   Achondroplasia
   Hydrocephaly
   Acrocephaly
   Microcephaly
   Other congenital anomalies

xi) *Synostoses of uncertain origin*
   Scaphocephaly and others

3   Inflammation of bone

All types of bone inflammation of whatever nature, or however produced, may be referred to under the general term *osteitis*. If the infection affects only the outer (cortical) bone, it is known as a *periostitis*. When the inner (cancellous) tissue is chiefly involved, it is called an *osteomyelitis*. These terms are useful in general descriptions of bone disease, but it should be realized that owing to the arrangement of the vascular supply in bone,

infection usually affects more than one part of the bone. In an osteitis, the bone may become considerably thickened owing to the formation of a new irregular layer, which encases the normal bone; and may be perforated by several openings that permit any discharge.

Sometimes it is possible to attribute the bone deformity to a specific disease such as tuberculosis or syphilis (described separately), whereas others must be diagnosed simply as osteitis, periostitis or osteomyelitis. One form of osteitis that is localized and yet quite distinctive is that affecting the mastoid process, and this develops from an inflammation of the middle ear *(otitis media)*. Cases of mastoid trouble have so far been noted mainly in Nubian, Egyptian and American material (Sigerist, 1951). An interesting case of this was found in an early Dynastic skull from Tarkhan, where the disease of the left mastoid is associated with a trephine hole on the same side (see Brothwell in Oakley *et al.*, 1959).

Periostitis may result from a blow, in which case bone thickening may be very limited. In more chronic states, *periostitis ossificans* develops. A good example of this is to be seen in a medieval fibula in the British Museum (Natural History) (Pl. 6.1). New bone has been produced on the exterior surface, and the normally smooth surface made rough and rugged. Numerous such cases of periostitis are known, and in French and British material, some specimens are as early as Neolithic (Pales, 1930; Brothwell, 1961b).

Osteomyelitis has also been frequent from Neolithic times onwards. It may develop as a result of bacteria getting directly into the bone at the time of a compound fracture, or through infection spreading there from another area of the body. It should be noted that osteomyelitis does not always affect the whole bone or shaft. The tibia of a Saxon skeleton from Thurgarton, England (Pl. 6.2), in fact shows acute osteomyelitis in the proximal end. Probably, bacteria were transported into the bone by the blood, and produced an abscess. A canal has formed to the exterior, in order to allow pus to escape, and there is evidence of an additional spread of infection at the opening. In a Medieval humerus from Scarborough (Pl. 6.2), it seems possible that a similar inflammation took place, but at a sufficiently early age for the epiphyses to be destroyed, with the result that the proximal end is badly deformed and the length of the bone much reduced.

4   Evidence of tuberculosis

This disease has caused much debate among palaeopathologists during the past half-century, as a result of which a number of the earlier claims (particularly of 'tubercular arthritis') are in doubt. However, we may provisionally accept the fact that the disease has been detected in some prehistoric and protohistoric specimens from Europe (Pales, 1930; Brothwell, 1961b). There is also some evidence that the disease occurred in Greek and Roman antiquity and in ancient Egypt (Sigerist, 1951, Morse, Brothwell & Ucko, 1964, Morse, 1967). The soft tissues of Egyptian mummies have provided the most concrete evidence that it afflicted early human populations. The classic example of tuberculous infection of the spine (called Pott's disease) is in a mummy of the priest of Ammon, where a large psoas abscess was found in association with vertebral destruction (Moodie, 1923).

Tuberculous osteitis may begin to develop in the end of a long-bone, or in one of the short bones of the body (such as the vertebra). There is destruction of bone, and eventually joints may become involved. It occurs most often in children, and in earlier times was more likely to have been fatal; but as children's bones are more liable to become eroded and to disintegrate, it is not surprising that relatively few cases of the disease have come to light in excavations. Kelly & El-

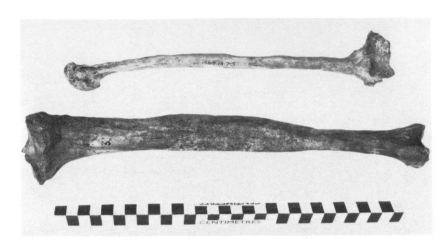

Plate 6.1
A.   Swollen tibia shaft
and wasted humerus in a
female Veddah from
Ceylon. It is very
probable that the
individual suffered from
yaws.

B.   The left and right
fibulae of a medieval
skeleton from
Scarborough. The right
shaft displays extra bone
formation resulting from
periostitis.

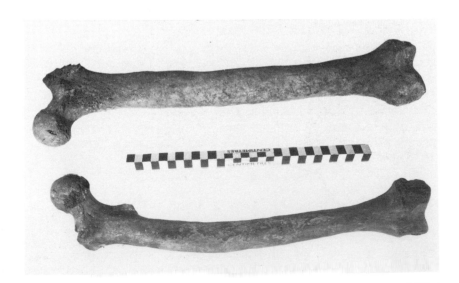

C.   Femur from a
17th–18th Century
Londoner, showing
marked shaft swelling
possibly due to syphilis.

D.   Femur from a
17th–18th Century
Londoner, with a bowed
shaft due to rickets.

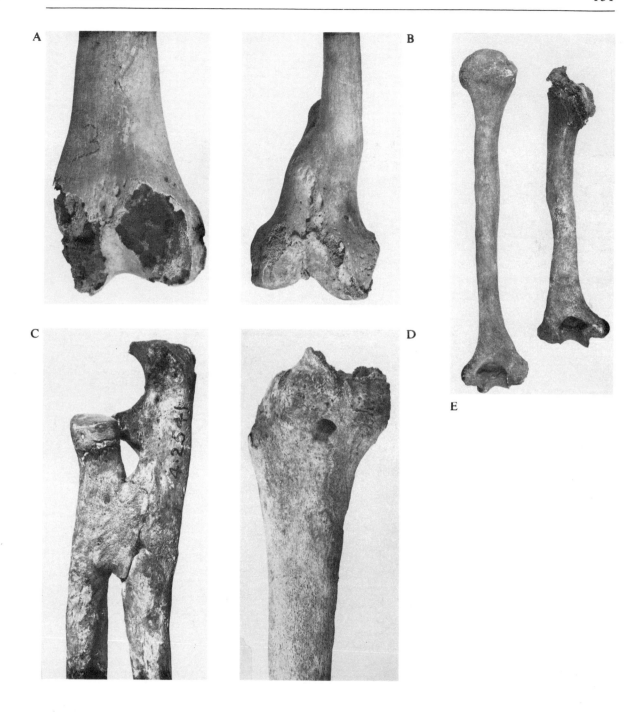

Plate 6.2   A, Lower part of a femur showing extensive damage by insects to the condylar region. From Socotra. B, Healed fracture of the femur, showing considerable deformity at the distal end of the shaft. Guanche female. C, United ulna and radius in the region of an old fracture. Saxon. Hants. D, Tibia with osteomyelitis of the proximal end. Note rounded sinus aperture. Saxon. Notts. E, Normal and abnormal humeri from medieval Scarborough skeletons. The humerus on the right shows complete destruction of the head, perhaps through childhood osteomyelitis in this region.

Najjar (1980) note that combinations of skeletal lesions can occur (spine–rib, spine–rib–sternum, spine–hip), and there may assist in diagnosis.

In the case of tuberculosis of the spine, the thoracic and lumbar regions are most commonly involved. At first the body of one vertebra softens and breaks down, and the disease may then spread to the next vertebra until a hump-backed condition (angular kyphosis) is produced. Some degree of repair may set in, resulting in the bony union of two or more vertebrae. In considering the differential diagnosis of spinal osteitis, one should also keep in mind the less likely alternatives of pyogenic osteitis, typhoid spondylitis, brucellar spondylitis, vertebral syphilis, and mycotic infection (Mantle, 1959).

To summarize the present position regarding the identification of tuberculosis in early bones, vertebral deformity is by far the most reliable diagnostic feature; other skeletal changes are far less conclusive. Occasionally, bone pathology may be associated with plaques of pleural calcification, as in three medieval Danish individuals (Weiss & Møller-Christensen, 1971).

A typical example of spinal tuberculosis has been described by Schaefer (1955) in an early female skeleton (eighth–eleventh century) from Germany. The woman must have been markedly humped-backed, for the vertebrae in the region of the tubercular infection had fused into a solid block. Evidence for tuberculosis of antiquity in America has been constantly sought over many years, and this is a matter of interest because until recently, hump-back representations in pottery were more common than actual cases in skeletal material. Indeed, Linné (1943) considered such effigies as indicative of tuberculosis (if not of rickets or syphilis) in pre-Spanish times. Possible evidence of tuberculosis in America has now been recorded by various authorities. Garcia Frias (1940) described three prehistoric mummies from Peru that showed spinal kyphosis, and Requena (1946) also diagnosed tuberculosis on the basis of three diseased dorsal vertebrae from pre-Spanish level in Venezuela. Ritchie (1952), described three kyphotic Indian spines from three separate prehistoric pre-Columbian cultures in New York State, and there is little room to doubt that these deformities were produced by tuberculosis (one of his cases is shown in Pl. 6.3). By far the most detailed study of possible tuberculosis of bone in Amerindian and other populations is that by Morse (1967). Buikstra (1977) and Buikstra & Cook (1978) have very usefully discussed the differential diagnosis of tuberculosis in a series of Pre-Columbian specimens, particularly in relation to another disease, blastomycosis. They demonstrate that there has been a tendency to underrate the occurrence and cultural implications of this kind of infectious disease in the past.

Perhaps it should be pointed out here that evidence of collapsed centra or vertebral bodies (Fig. 4.18) need not necessarily imply tuberculosis. Stewart (1956a) has found such abnormalities in Eskimo skeletal material (Fig. 4.18), and is inclined to think that some at least may have been caused by compression fractures. Gaspardy & Nemeskéri (1960) describe another case of a compression fracture in a Copper Age skeleton from Hungary, and I have noted less severe instances in early British specimens.

One should also be aware of the possibility of other lesions of the vertebral body. Brucellosis osteitis, for instance, can result in inflammatory changes, and this may have occurred in early Jericho (Brothwell, 1965). Wells (1974) notes restricted destruction of two lumbar vertebrae in a Bronze Age individual, due to adolescent osteochondritis.

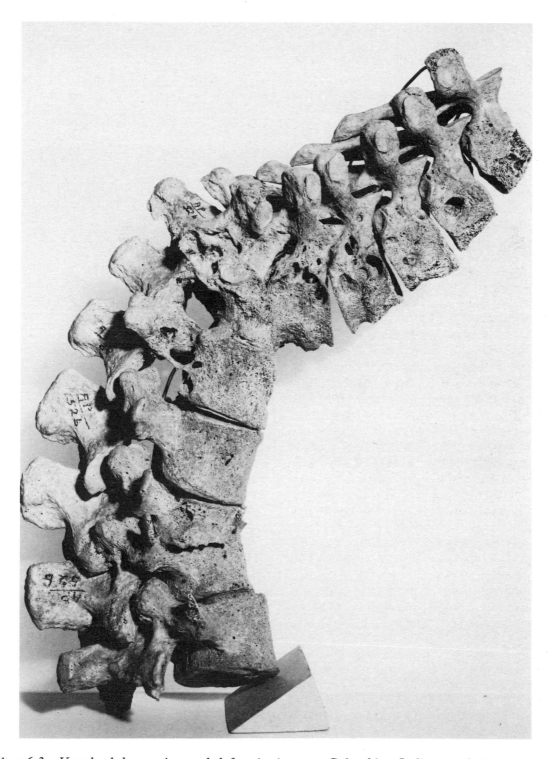

Plate 6.3   Vertebral destruction and deformity in a pre-Columbian Indian, probably caused by tuberculosis. (By courtesy of Dr W. A. Ritchie.)

5   Evidence of treponematoses

The treponematoses

One genus of spirochaete micro-organism called *Treponema* has considerable interest ot those working in the field of palaeo-epidemiology. This group of diseases includes the modern clinical conditions known as pinta, yaws, endemic syphilis and venereal syphilis (Hackett, 1963). The clinical expression of these diseases is seen in Table 7. It could well be that there has been a long association of the treponemes with the hominids, and thus it is more than likely that their relationship with man has changed to some extent through time. Modern clinical names can thus only be used in a tentative way in relation to possible ancient treponematoses (the range of diseases caused by *Treponema*).

It can be seen in Table 7 that three of the treponemal diseases produce bone changes, at least in more advanced stages. Although this bone pathology can at times mimic other disease, Hackett (1976) shows clearly the sequence of bone changes that usually occur (Figs. 6.1–6.4). It will be seen that the bone changes to the skull and long-bones are particularly important in the diagnosis of treponematosis. In the case of the skull, and in marked contrast to bone destruction resulting from leprosy or secondary tumours (metastases), there is usually a widespread vault (and to a lesser extent facial) osteitis, beginning with restricted zones of periostitis (the so-called 'clustered pits' of early inflammation) leading to extensive deep regional 'cavitation'. Bone remodelling, with radial scarring, can occur in the later stages.

In the case of the long-bones, the sequence of changes ranges from early sub periosteal changes in the form of pitting, striations and restricted extra bone through better-defined nodes and 'superficial cavitation' to late stage highly modified swollen cortical tissue (less dense) with cavitation. In late treponemal osteitis of the tibia, there can be marked

bowing ('sabre' tibia), a feature particularly seen in Australasia (Hackett, 1967). As yet, no definite case of syphilis has been established in prehistoric European material (Sigerist, 1951) although possible medieval cases are known (Pl. 6.1). Many American bones, possibly of both pre- and post-Columbian date, have been described as syphilitic, and there is no doubt that some of them could be. Goldstein (1957), for instance, found possible evidence of this disease in

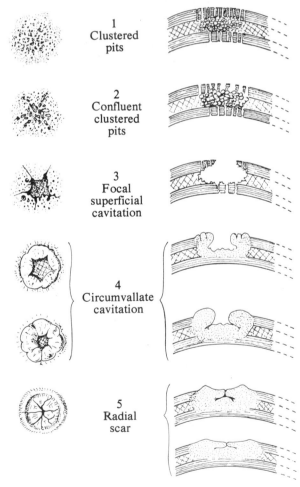

Figure 6.1   Syphilis sequence in skull bones.
(Redrawn from Hackett.)

early Texan material, both on skull and long-bone fragments. Stewart (1956*b*), although critical of the claims of pre-Columbian syphilis in the Americas, described human remains from a ninth–tenth century site in Central Mexico, and in one skeleton the tibiae showed bowing and extensive disease changes.

Evidence of yaws

Yaws and syphilis, whether one is considering bony changes or the microscopic organisms themselves, are notoriously difficult to distinguish. Thus although few cases of yaws have been diagnosed in archaeological material, it seems worth discussing them in some detail. The early history of this disease is obscure, until the seventeenth century when accounts of it came from Brazil, the East

Indies and West Indies (Clemow, 1903). Later it was recognized in Africa and in some of the Pacific Islands. Nomenclature pertaining to it is still very confusing, as one survey shows (Hill, Kodijat & Sardini, 1951), and thus bone changes will be discussed without reference to the stage of the disease.

Unlike veneral syphilis, infection often begins during childhood. If bones become affected, localized *gummata* (nodes) may be present, or a more diffuse osteitis. In the former case, irregular depressions in the bone may result, while the latter gives rise to bowing and deformity, especially in the tibia, fibula and ulna (Spittel, 1923). The bones may show thickening and greater density both externally and internally, although small circular areas of rarefaction may also be associated with these changes. According to Spittel, yaws affects bones more extensively

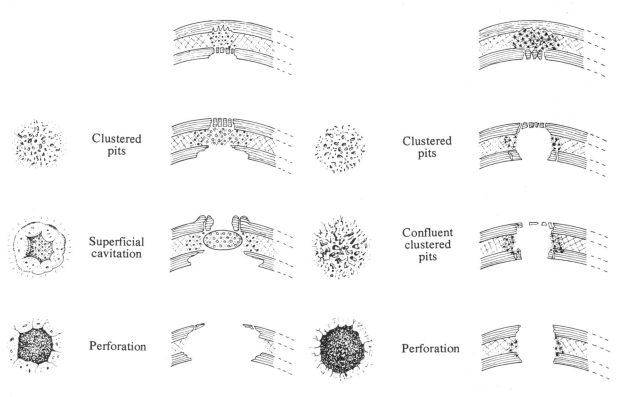

Figure 6.2 Tuberculosis sequence in skull bones. (Redrawn from Hackett.)

Figure 6.3 Metastatic neoplasm sequence for comparison with treponemal changes of the skull. (Redrawn from Hackett.)

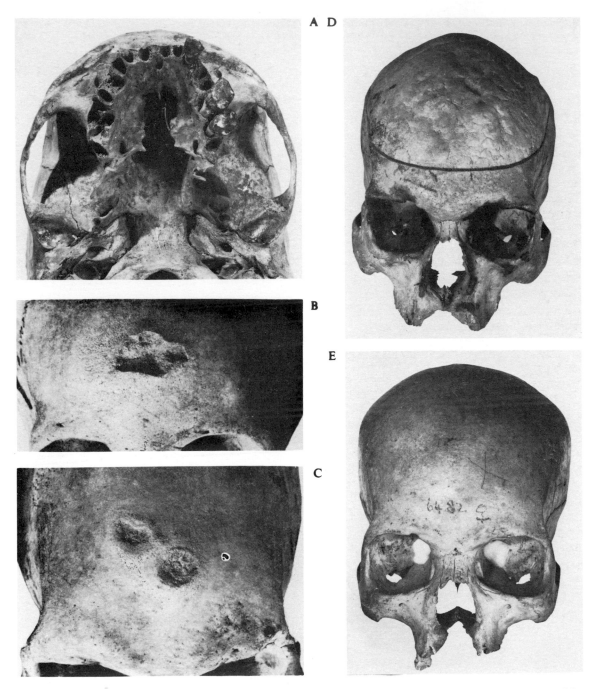

Plate 6.4   A, Palatal region of a female Veddah with considerable palatal destruction caused by yaws. B, Prehistoric skull from the Western Pacific showing a crater-like lesion possibly due to yaws. (By courtesy of Dr T. D. Stewart and Dr A. Spoehr.) C, Frontal region of an Australian aboriginal showing similar lesions. (By courtesy of Dr T. D. Stewart and Dr A. Spoehr.) D, An Australian aboriginal skull with extensive infection and destruction of the vault and palatal regions, probably through yaws. E, Skull of a Gaboon Negro with considerable palatal destruction. Diagnosis in this case is difficult to establish.

Table 7

*Some clinical characters of the Treponematoses**

Character	Pinta	Yaws	Endemic syphilis	Venereal syphilis
Usual source of treponemes	Skin anywhere	Skin anywhere	Buccal mucosa	Genital and mucosal lesions
Size of infectious area	Large	Large	Small	Small
Duration of:				
Infectiousness of individual lesions	Many years	A few months	A few months	A few months
Infectiousness of patients, including infectious relapses	Many years	3–5 years	3–5 years	3–5 years
Latency[a]	Absent	Characteristic	Characteristic	Characteristic
Lesions:				
Initial, site	Exposed skin	Skin of legs	Mouth?	Genital
Initial, occurrence	Infrequent	Frequent	Unusual	Frequent
Generalized skin, extent	Extensive	Extensive	Limited	Moderate
Genital, occurrence	Unusual	Scanty	Scanty	Frequent
Buccal mocosal, occurrence	Absent	Scanty or absent	Moderate	Scanty
Palmar and plantar hyperkerafoses, occurrence	Absent[b]	Frequent	Scanty[c]	Scanty
Bone, occurrence	Absent	Moderate	Moderate	Scanty
Juxta-articular nodules, occurrence	Absent	Frequent	Scanty	Scanty
Heart, brain and other viscera, occurrence	Absent	Absent[d]	Scanty and mild, or absent	Moderate
Congenital transmission	Absent	Absent	Absent	Present
Age-group of most infections	Children	Children	Children	Adults

[a] Latency in a communicable disease is a stage in its course in which there is evidence of infection such as seroreactivity but no clinically detectable active lesions.

[b] Hyperkeratoses have been attributed to pinta in Cuba. However, there may have been confusion with hyperkeratoses of yaws.

[c] Many of the hyperkeratoses illustrated by Hudson (1936) and Murray *et al.* (1956) are not due to endemic syphilis (Hackett & Loewenthal, 1960).

[d] Weller (1937) reports lesions resembling those of venereal sylphilis from supposed yaws cases in Haiti; he found no venereal syphilitic lesions of the brain in them.

* Hackett, 1963.

than syphilis, and in severe cases nearly all the bones of the extremities may be diseased.

On the head and facial bones, characteristic depressed scars may be found, and in more advanced stages, the hard palate may be completely destroyed and the nasal region generally attacked. Even primitive groups today recognize the horrible nature of this facial destruction, and a number of masks clearly portray the disease (as figured by Simmons, 1957). The British Museum (Natural History) has a number of specimens believed to be affected by yaws, but all of recent date. One female Veddah skeleton, first examined by Osman Hill (1941), shows bone destruction in the nasal and palatal region (Pl. 6.4), as well as minor swellings on some long-bones. It is also of particular interest in showing marked thinning of the

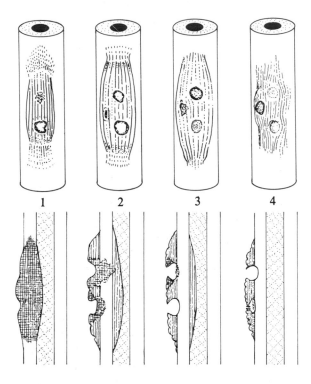

Figure 6.4   Syphilis sequence in long-bones.
(Redrawn from Hackett.)

humeri (Pl. 6.1), which may be attributable to deformity, joint fixation, and resulting atrophy through disuse. The possibility that the disease in this person is yaws has been confirmed by Spittel (1957) in a personal communication.

More certain evidence of yaws has been given by Stewart & Spoehr (1952) in a study of bones from a prehistoric site on the Island of Tinian in the Western Pacific. Thickening was noted in a number of long-bones, one tibia shows some degree of bowing, and a number of pits were also to be seen. The skull (Pl. 6.4) shows a number of crater-like lesions that are unmistakably similar to bone lesions found in modern cases of yaws. Surprisingly, the nasal region appears to have escaped without deformity.

Further possible evidence of this disease group in the New World is provided by various bone pathology to be seen in Amerindian skeletons from Florida (Bullen, 1972).

The history of venereal syphilis, and the related treponemal diseases (endemic syphilis, yaws and pinta), have been discussed by Hudson (1946), Hackett (1963), Brothwell (1970, 1978) and others. It is a complex problem with nearly as many views as skeletons to support them!

6   Evidence of leprosy

There is no doubt that leprosy was one of the most deforming and most publicized diseases in early historic times. It was a particularly familiar disorder in Medieval Europe and considerable antiquity has been claimed for it in India and China (Clemow 1903).

A special report of the World Health Organization (1953) recognizes four main forms or classes of leprosy. Of these, it is possible that 'indeterminate (maculo-anaesthetic) leprosy' shows the greatest frequency of associated bone changes.

Although various studies, particularly of a radiographic nature, have discussed bone changes in modern lepers, nothing was known of leprosy in earlier man until Møller-

Christensen's (1953) study of skeletons from a Medieval Danish leper hospital. As it seems highly probable that the changes noted in skeletons from this burial ground will be present in early leper bones from Britain and other parts of Europe, the principal bone changes may be listed as follows (Møller-Christensen, 1961; Andersen, 1969). Bony changes do not always accompany leprosy, of course, nor are they present to the same degree in each case; thus the detection of leprous changes in earlier skeletons gives only a rough idea of the frequency of the disease among the living.

The skull

(See Pl. 6.5)

a) Specific atrophy of the alveolar bone in the region of the upper incisors (with or without resulting loss of teeth).

b) The hard palate may show minor osteitis or areas may be completely resorbed.

c) The anterior nasal spine may be absent or considerably reduced, and is often associated with atrophy of the margins of the pyriform aperture.

d) Danielsen (1970) provides evidence that low-resistance leprosy in childhood can at times produce tooth malformations, especially in root growth (odontodysplasia leprosa).

The post-cranial skeleton

a) The phalanges of the hand may show enlarged nutrient foramina, and various degrees of degenerative reduction in size.

b) The tibia and fibula may show well-marked vascular grooves, with or without an associated mild periostitis of the shaft.

c) The metatarsals and phalanges may show deformity and bone destruction (Pl. 6.5). There may also be enlarged nutrient foramina, and secondary arthritic changes may result in fusion of bones.

It is believed that leprosy was brought to Britain during the Roman invasions, and leper-houses were established in Saxon times. As regards leprosy in British skeletons, there are a few possible cases. A medieval skull from Scarborough (Brothwell, 1958b) now in the British Museum (Natural History), shows changes in the nasal region that could have been caused by this discase. A seventh-century skeleton excavated in the Isles of Scilly shows typical leprous changes in the skull and post-cranial bones (Brothwell, 1961b), and a Saxon skeleton from Beckford displays characteristic changes at the feet (Wells, 1962). Finally, there is now clear evidence for the movement of leprosy through into northern Europe by late Roman times, in a fourth century A.D. skeleton from Poundbury Camp, Dorchester (Reader, 1974).

7   Bone tumours

Tumours causing bone changes, especially malignant cases, were uncommon in earlier man, but possibly not as rare as some have previously supposed (Brothwell, 1967a). None of the tori of the skull are discussed here, as it seems doubtful whether they can be classified as tumours. Care must be taken when assigning 'bumps' on the post-cranial bones to this class of disease, for as Sigerist (1951) says, it is sometimes difficult to decide whether an exostosis (warty outgrowth) is a tumour or the end result of an inflammatory process. For example, it is possible that the bony projection on the first-discovered femur of *Homo erectus*, discovered by Dubois, is an outgrowth resulting from an injury to the

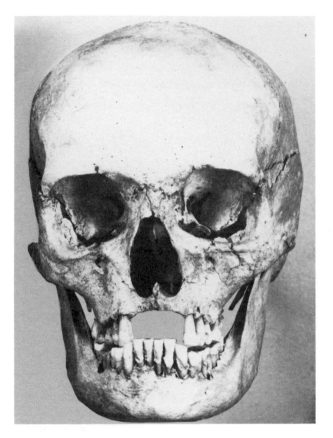

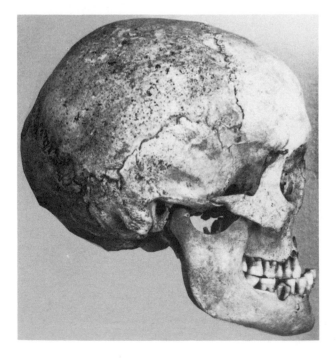

Plate 6.5   Bone changes in leprosy. The skull and feet from a Danish medieval skeleton (Case I from Naestved). (By courtesy of Dr V. Møller-Christensen.)

shaft (Fig. 6.5) and not a malignant tumour. On the other hand, the so-called 'chin' of the supposedly Lower Pleistocene Kanam jaw may in fact be largely a bony reaction to a 'sub-periosteal ossifying sarcoma' (Tobias, 1960).

A review of the evidence for tumours in early British man has been published, and for a description of these the reader is referred to Brothwell (1961b). A far more expanded review of tumours than that given below is undertaken in Brothwell (1967a). It is emphasized that the differential diagnosis of such diseases is far from an easy matter, and for instance in the multiple destructive lesions found in an early Egyptian skeleton from Assyut, Satinoff (1972) was not able to suggest the area for the primary tumour, simply remarking on the secondary deposits.

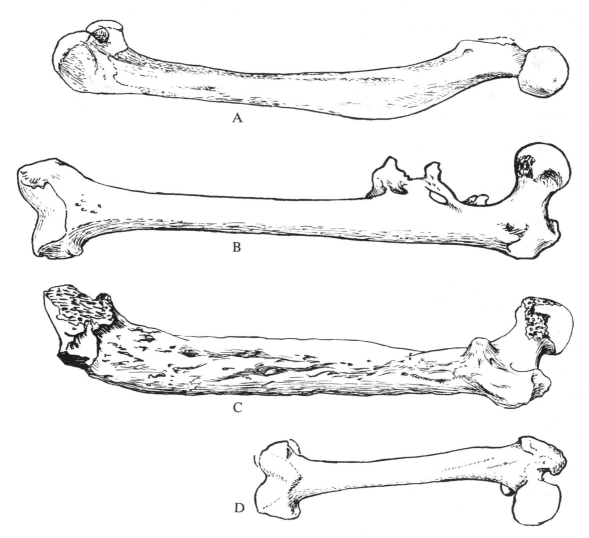

Figure 6.5   A, deformed (bowed) femoral shaft caused by rickets. Seventeenth to eighteenth century Londoner. B, an extensive medial exostosis possibly due to trauma in the first Homo erectus *femur. C, swollen femur shaft from a French Neolithic skeleton, possibly due to Paget's disease. D, femur of an achondroplastic dwarf from the Royal Tombs, Abydos (Early Dynastic).*

i) Simple tumours

Osteomata

a) *Ivory or benign osteomata.* This is by far the commonest form of tumour to be found in archaeological material. The tumours are seen as simple mounds of compact bone, usually on the external aspect of the cranial vault (Fig. 6.6). They have been noted in British material from Neolithic times onwards. Rarely is there more than one mound. Osteomata may also develop in the orbital cavity or in one of the air sinuses. Goodman & Morant (1940) noted an orbital

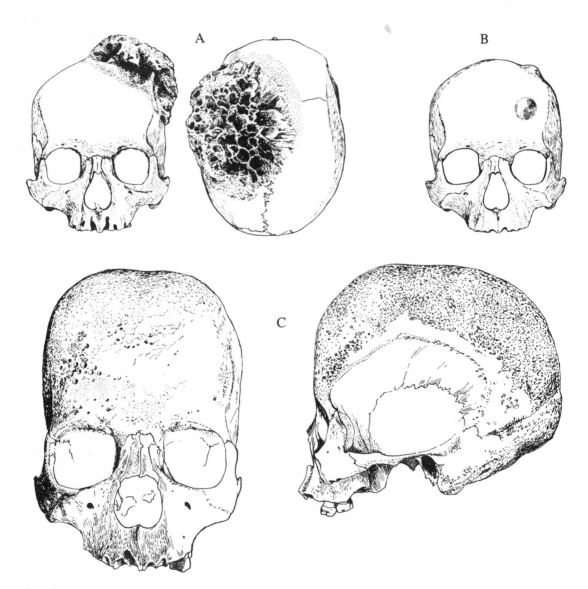

Figure 6.6   A, facial and top views of a Peruvian skull, showing an extensive osteosarcoma on the left side. B, typical examples of benign (buttons) osteoma. Side and top views are illustrated. C, facial and lateral views of a Nepalese skull, showing considerable thickening of the frontal and parietal bones probably due to a blood disorder.

tumour, probably of this kind, in material from Maiden Castle, Dorset.

b) Other osteomata. Simple tumours that are made up of inner cancellous tissues as well as compact bone may be of this category. Goldstein (1957) notes a large tumour of the humerus in an early Texan Indian, which appears to be an osteoma of this kind.

ii) Malignant disease of bone

a) Osteosarcoma

The growth may be ingrowing or may extend on the original surface of the bone, but only the former condition has so far been found in archaeological specimens. Cases of possible osteosarcoma or osteochondroma have been described in Egyptian long-bones of the Fifth Dynasty by Elliot Smith & Dawson (1924). Ruffer (1920) reported another tumour, probably of the same kind, in a pelvis from the ancient catacombs of Kon el-Shougata in Alexandria. Hug (1956) has also recorded the presence of a tumour (apparently an osteosarcoma) in an Iron Age skeleton from Münsingen (Pl. 6.7). Of a more recent age is the tumour described by MacCurdy (1923) in a pre-Columbian Peruvian skull (Fig. 6.6).

b) Multiple myeloma

This disease is usually characterized by the formation of multiple tumours. These take the form of clear rounded lesions, generally not more than a centimetre in diameter, although occasionally much larger. The bones most frequently affected are the spine, ribs, skull, femora, clavicles and humeri. Punched-out holes are produced (Pl. 6.7), especially in the skull, the cortex (inner part) of the bone being eroded from within. In modern man, its occurrence is mainly restricted to the age period 40 to 60 years.

Fusté (1955) has recorded the presence of such lesions in a late Neolithic ('Neo-eneolithic') skull from the Pyrenees. Ackerknecht (1953) also recalls that multiple myeloma have been reported in Neolithic French and pre-Columbian North American material. Ritchie & Warren (1932) and more recently Brooks and Melbye (1967) describe typical cases in pre-Columbian Indians. It should be remembered that post-mortem corrosion may sometimes produce rather similar holes. Also, secondary tumours following a primary cancer may also simulate myelomatosis, and it could well be that some of the cases claimed to be the result of myeloma formation may turn out to be metastases (secondaries).

8   Diseases of joints (arthritis)

It is perhaps, a little consoling to learn that joint diseases that aggravate and cripple so many modern human beings, especially in the civilized parts of the world, are known to have occurred even in Pleistocene man. Evidence of some form of arthropathy is present in more than one Neanderthaler, and the well-known skeleton from La Chapelle-aux-Saints shows quite severe bone changes on the vertebrae and at the mandibular condyles (Straus & Cave, 1957). Cro-Magnon man was not less affected (Ackerknecht, 1953), and by Neolithic times, numerous cases may be cited. Indeed Elliot Smith & Dawson (1924) considered arthritis to be the chief disease of ancient Egyptians and Nubians. It seems probable that certain areas are involved more frequently in some groups than in other (Bourke, 1967). In many specimens, the degree of development of joint changes suggests a restricted working capacity on the part of the individual, and it is this aspect of disablement in a primitive community that may eventually yield the

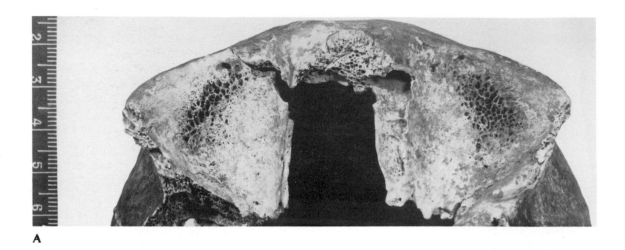

A

B

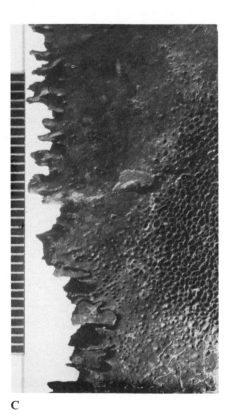

C

Plate 6.6   A, The frontal aspect of orbits displaying a medium degree of osteoporosis (usura orbitae). B, The vault of a late Chalcolithic female from Iran, showing extensive bony changes and thinning due to a purulent scalp infection. C, Part of the parietal of a late Saxon child from Nottinghamshire, showing a medium degree of osteoporosis.

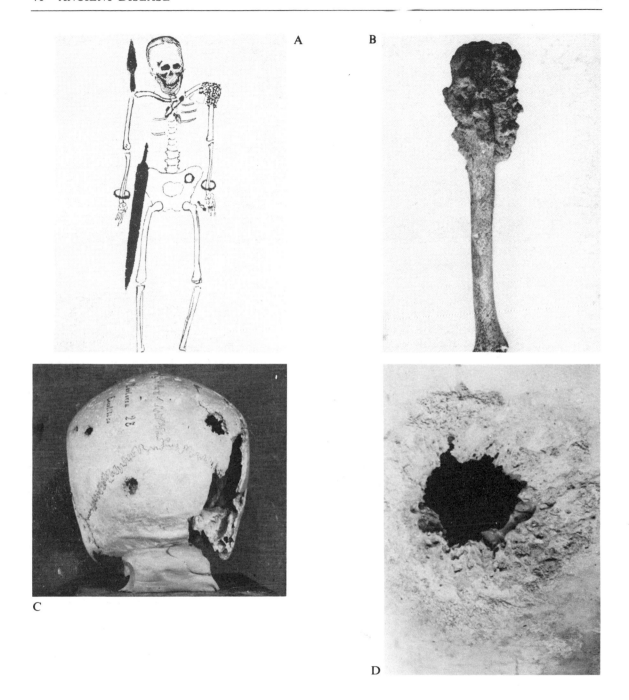

Plate 6.7   A, An Iron Age skeleton from Munsingen showing an osteosarcoma at the left shoulder. (By courtesy of the Naturhistorisches Museum, Bern.) B, The humerus of this skeleton. (By courtesy of the Naturhistorisches Museum, Bern.) C, 'Neo-eneolithic' skull from the Pyrenees, displaying perforations caused by multiple myeloma. (By courtesy of Dr M. Fusté.) D, The same skull with a perforation in close-up. (By courtesy of Dr M. Fusté.)

most interesting information from an archaeological point of view.

Contrary to popular belief this disease group is not restricted to cold and damp climates, but is found in different groups and environments throughout the world. The causes of these joint diseases are still far from being completely understood. Trauma (damage) in bone explains some cases. Certain diseases, dietary disturbances and severe working conditions may all be contributory factors in some instances, but to what extent it is difficult to tell. Age is certainly an important factor and some changes may be the result of 'wear and tear' through increasing body weight and faulty spine movement (Law, 1950). Mechanical stress can be very localized and yet produce a commonly occurring sequence of well-defined joint changes, as Ortner (1968) found in his study of the distal end of the humerus.

As yet, the several varieties of arthropathies have not been fully distinguished in the skeleton (Stewart, 1950), and in this discussion, therefore, only four main categories will be outlined (the divisions being a modification of Bourke, 1967).

These are as follows:
- a) Acute suppurative arthritis
- b) Chronic arthropathies
    - i) Rheumatoid arthritis
    - ii) Osteoarthrosis
    - iii) Ankylosing spondylitis
- c) Gout
- d) Other arthropathies

a) Acute suppurative arthritis

Knee and hip joints are most commonly affected by non-tubercular organisms. Brucellosis and the typhoid bacillus sometimes produce joint changes (especially of the vertebrae). Syphilis can also damage the joints. A good example of this category is probably the adult male for Tula in Mexico (Db1–2), who shows one hip joint severely modified by an inflammatory condition (Dávalos, 1970). A further example, in a humerus from medieval Scarborough, is seen in Plate 6.8. Changes in vertebrae that are possibly the result of brucellosis are reported in a Bronze Age individual from Jericho (Brothwell, 1965).

b) Chronic arthropathies

i) Rheumatoid arthritis

This disease may show no clear cause, or may be associated with a definite infection, and the onset is generally between twenty and forty years. Associated changes include narrowing of the joint space, bony 'lipping', and more occasionally, ankylosis (fusion). The joints of the hands and feet are most commonly affected (Pl. 6.8), but other joints can be involved. It may well be significant that very few possible cases of this condition have been reported in archaeological material. One likely case has occurred in a seventh century A.D. skeleton from the Isles of Scilly (Brothwell & Møller-Christensen, 1963).

ii) Osteoarthritis (osteoarthrosis, osteophytosis, spondylarthrosis).

Judging by the usual positions of bony changes in earlier skeletal material, it would seem that this type of arthropathy was far more common (though the frequency may be distorted through the smaller hand and foot bones usually receiving only a rather casual examination). As in rheumatoid arthritis, the cause may be completely unknown, or directly related to injury, infection or more probably age-related stress and strain. The disease usually affects middle-aged or older individuals. Bony 'lipping' (known also as 'osteophytosis') occurs in every case (Pl. 6.9). Narrowing of the joint space is present in about half the cases, but except in the spine, ankylosis is rare. One or more of the larger joints are commonly affected, and a vertebral joint is shown in Fig. 6.7. As an example of osteoarthritic changes to be seen in long-bones, the knee of a medieval man from Scarborough may be described. In this specimen (Pl. 6.8) there is considerable

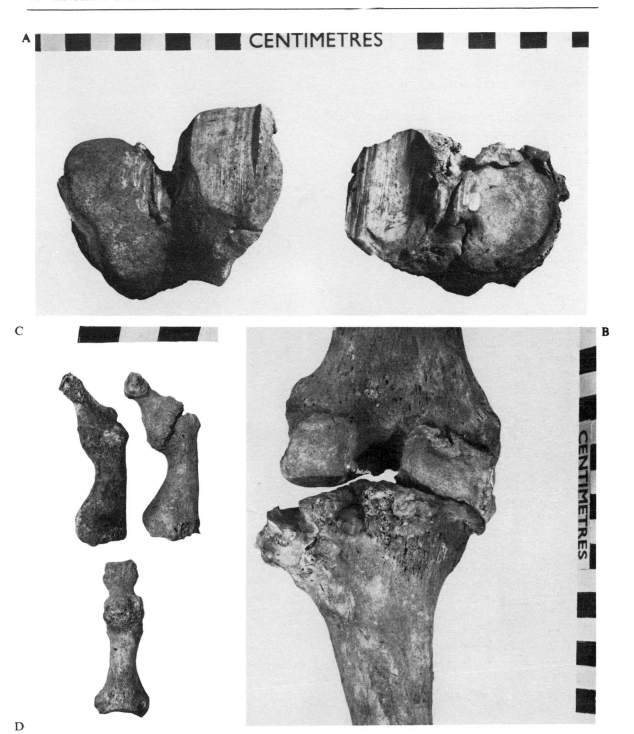

Plate 6.8   Rheumatic changes in bones. A–B, Severe osteoarthritis in a medieval male from Scarborough. A, Articular surfaces of the femur and tibia (at the knee) showing osteoarthritic lipping and considerable eburnation (through the movement of bone on bone). B, Posterior aspect of the knee showing 'lipped' margins of the condyles. C–D, Fused phalanges resulting from arthritis.

lipping at the margins of the condyles in both tibia and femur, with eburnation (the 'ivory' polish sometimes produced by bone moving on bone) in one area where the cartilage had disintegrated during life.

The hip joint is one of the most affected areas. Lipping and roughening may develop

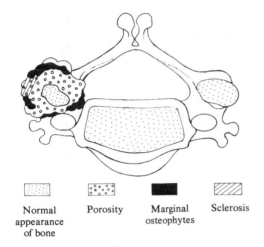

Figure 6.7   Lateral joint surface with moderately severe spondylarthrosis. The area of this joint surface is about twice the size of the normal joint surface on the opposite side (after Sager, 1969).

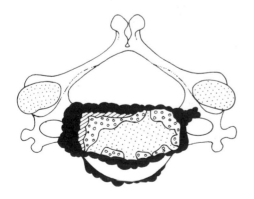

Figure 6.8   Diagram of the macroscopic bone changes seen in the discal surface of the cervical body in moderately severe degenerative changes (after Sager, 1969).

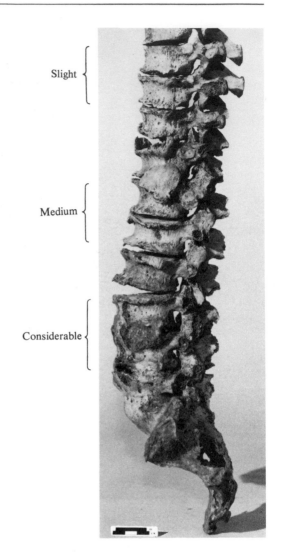

Plate 6.9   The vertebral column of a premedieval skeleton from Cambridge, showing various degrees of osteoarthritic 'lipping'.

at the outer margin of the acetabulum and it is usual to find distortion of the femoral head, especially where it joins the neck of the femur (Pl. 6.10). Eburnation may be present on the head.

By far the commonest area to be involved is the spinal column (where the arthritic condition is sometimes mistakenly referred to as spondylitis, the rarer and more severe joint condition). Although bony lipping and spur

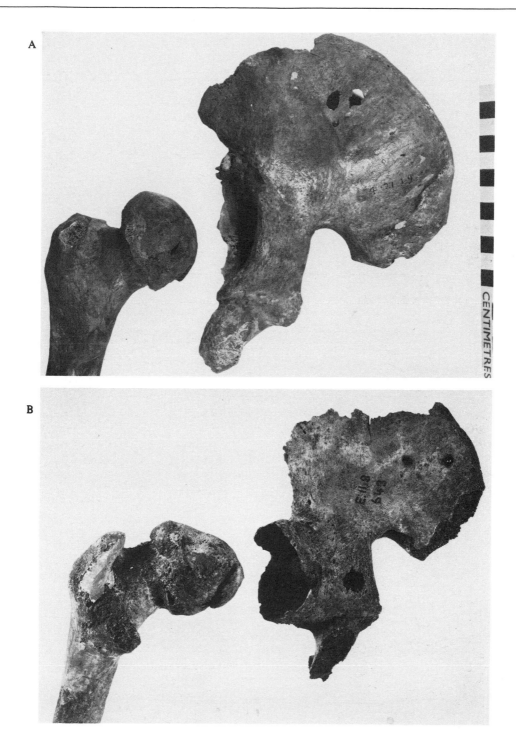

Plate 6.10   Abnormalities of the hip. A, Abnormally shallow acetabulum and deformed femoral head, possibly indicative of a partial dislocation. Anglo-Saxon. Guildown. B, Severe osteoarthritis of the hip in a Romano-Britain. Note the very 'lipped' acetabulum and deformed femur head.

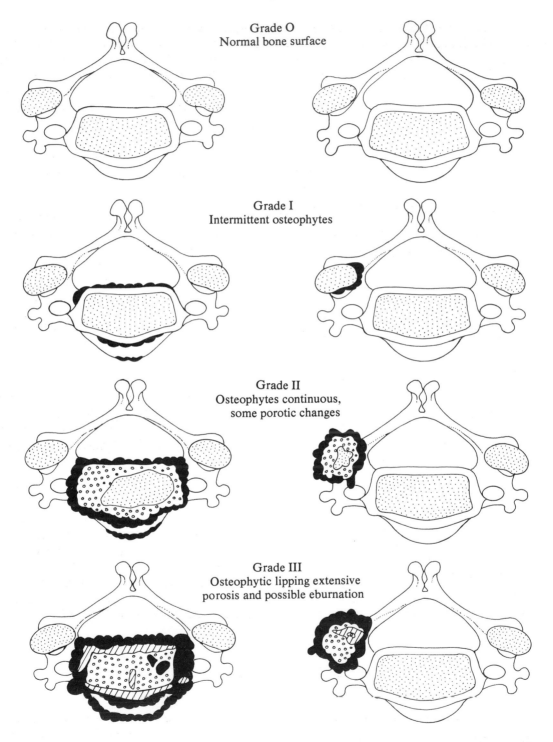

Grade O
Normal bone surface

Grade I
Intermittent osteophytes

Grade II
Osteophytes continuous,
some porotic changes

Grade III
Osteophytic lipping extensive
porosis and possible eburnation

Figure 6.9   Left, schematic drawings of the macroscopic bone changes seen on the discal surface of cervical bodies in progressive grades of degeneration. Right, joint surface diagrams of cervical vertebrae with progressive grades of lateral spondylarthrosis (after Sager, 1969). See Figure 6.7 for key.

formation may be found at the articular surfaces and transverse spine, it is more usual to find only the vertebral body noticeably involved with less marked degrees of change elsewhere (Fig. 6.8). Both the dorsal and ventral margins of the body may show some degree of lipping. Some idea of the progressive nature of the vertebral joint changes is seen in Fig. 6.9, from a detailed study by Sager (1969). Plate 6.9 shows an early British vertebral column in which all degrees of vertebral body osteoarthritis are shown, including complete fusion of the lower lumbar vertebrae with the sacrum. A fairly recent study by Stewart (1958b) shows that arthritic lipping progressively develops from young adult life, although more extreme degrees are not seen until after about 35 years. There is clearly much value in scoring precisely these osteoarthritic changes on each individual joint surface. From such careful work, one can begin to show possible changes between early populations in this respect. These data can be presented in various ways for comparative purposes, as shown by the work of Stewart (1966) and Bourke (1967).

Osteoarthritis of the temporo-mandibular joint was found in Neanderthal man at Krapina and at La Chapelle-aux-Saints. In this and later cases the main features are a flattening and roughening of the glenoid fossa and condyle of the mandible. Causes for this condition no doubt include severe dental wear, rigorous mastication and sometimes even malocclusion.

iii) *Ankylosing spondylitis*
This is a disease that more frequently affects males. Vertebral and sacro-iliac joints are involved in progressive changes (Riley, Ansell & Bywaters, 1971), and ossification also occurs in the spinal ligaments. Subsequent ossification may produce a fused vertebral block or 'bamboo spine'. Good examples of this condition are known in various cultures, especially Egyptian series (Bourke, 1967; Nielsen, 1970). On the other hand, Stewart (1966) notes its rarity in early Amerindians.

c) *Gout*
Although in mummified material, evidence of gout has been found in the form of multiple joint tophi and ulceration of the skin over the tarsal bones (Elliot Smith & Dawson, 1924); in inhumations one should be looking for the possible association of punched-out or irregular small joints of the hands and feet, with deposits of chalky material (which should reveal uric acid on analysis).

d) *Other arthropathies*

i) Charcot's arthropathy. A severe joint condition usually affecting the large joints of the limbs. A possible example of this condition is seen in a First Dynasty female from Abydos (Fig. 6.18A).

ii) Osteochondritis dissecans results in somewhat puzzling bone changes, namely, a punched-out lesion due to avascular necrosis occurring in the sub-chondral bone of a joint and subsequent degenerative changes. The joint concavities that result (Pl. 6.11) are usually under 20 mm in width and no more than 2–5 mm deep. Probably 80 per cent of the cases are found at the knee (Wells, 1974).

9   Diseases of the jaws and teeth

i) Dental caries and pre-mortem tooth loss

Probably most of those who read this handbook will have suffered to some extent from oral diseases, and in particular decay of the teeth. Contrary to earlier beliefs, it now seems possible that caries has always been associated with *Homo* ever since the genus became differentiated, and its presence in modern wild anthropoid apes (Colyer, 1936; Schultz, 1956) suggests that it could have affected earlier forms. Of the early Pleistocene hominids, caries has been noted in the Australopithecines of South Africa (Robinson, 1952; Clement, 1956), and in one of the Javanese varieties of *Homo erectus* (Brodrick, 1948).

The Neanderthal forms of man, existing in early Upper Pleistocene times, also displayed a tendency to caries, and Sognnaes (1956) noted a possible cavity in a specimen from Mount Carmel, Palestine. Perhaps the most surprising example of dental decay to be found in Pleistocene man is to be seen in the Rhodesian skull, where of 13 teeth (excluding 2 broken ones), 11 were carious (with 15 cavities in all). Carious teeth have also been noted in Aurignacian crania (Krogman, 1938*b*) and in two skulls from French

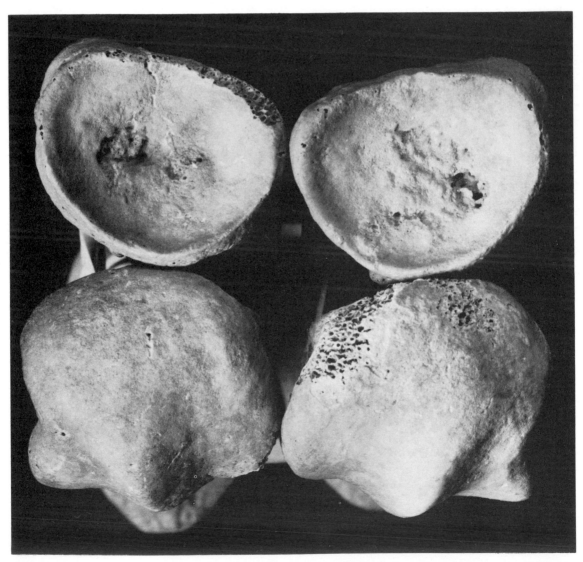

Plate 6.11   Pitting of the articular surface characteristic of osteochondritis desiccans in phalange bones from Tean, Dark Age skeleton.

Solutrean deposits (Vallois, 1936).

The frequency of dental disease may already have been on the increase by Mesolithic times, for of 24 skulls from Ofnet, Bavaria, 10 had caries and 3.8 per cent of the Mesolithic teeth from Teviec were also diseased. However, compared with later incidences of caries these figures are still extremely small, and presumably the hunting, fishing and gathering economy of these earlier peoples was beneficial to their oral health. There is a growing literature on the oral pathology of more recent populations (Miles, 1969; Pfeiffer, 1979).

In Europe, at least, significant increases in the frequency of dental decay probably began during the 'Neolithic Revolution'. Figs. 6.10 and 6.11 give some idea of the increase in caries and pre-mortem tooth loss from the Neolithic period onwards. Although there was in general an increase in dental caries, there would sometimes appear to have been slight decreases in caries, possibly associated with immigrant movements and dietary changes. The sharp increase in the frequency of caries in Britain after Saxon times is probably correlated with the introduction of more refined sugars and flours. The figure for tooth loss shows a similar picture, except for recent times, when the frequency may not show such a great increase.

Carious cavities usually develop in three regions of the tooth (Fig. 6.12):

a) On the occlusal (biting) surface, generally in the region of the natural fissures. This area is far less affected in earlier groups than in modern civilized ones.

b) In the region of the neck (cervical area) of the tooth, either on the lingual (tongue) or labial (lip) side.

c) In the region of the neck of the tooth, but between the teeth (mesial and distal).

When recording the presence of caries, it is advisable to note the area and specific teeth

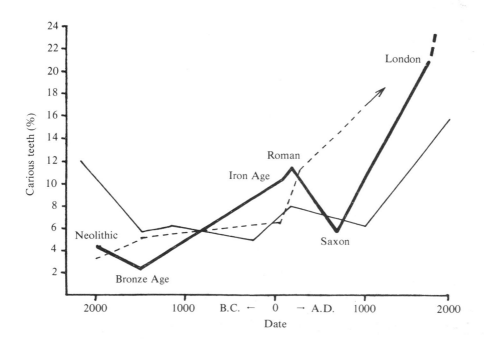

Figure 6.10   Caries frequencies through time in three areas. Mainly of young adults of both sexes. Britain: ▬; Greece: —; France: – – –.

affected. It is usual to find the molar teeth affected more frequently than the front teeth. It should be noted that an artificial non-carious interproximal grooving (Fig. 6.13) can occur at times (Brothwell, 1963; Ubelaker, Phenice & Bass, 1969). Although not common, it should be looked for and scored separately.

ii) Periodontal disease

The two most important affections influencing tooth loss are caries and periodontal disease, the latter probably being quite important in earlier man. There are a number of complicating factors, and one must keep in mind that extractions have been practised for thousands of years, as we know from ancient Mesopotamian records. Also, tooth evulsion for ritual reasons is undertaken by some primitive peoples, and even in Stone Age North African groups there is evidence of this procedure (Briggs, 1955).

Periodontal disease (sometimes known as Pyorrhoea) is an infection not only of the alveolar bone, but also of the soft tissues of the mouth, and any estimates of its frequency in skulls may be smaller than actually occurring, owing to the fact that minor degrees of infection may only involve the soft tissues. The effect of this disease on the alveolar bone is to cause its recession, and various degrees may therefore be noted (Fig. 6.14), resulting finally in the considerable loosening of the teeth and their loss. As in caries, various factors are involved in the production of this condition, including unclean mouths, irritation by calculus (tartar) deposits, attrition and lowered tissue resistance through a faulty diet.

A preliminary survey of the degree of alveolar recession in 130 skulls ranging in age from Neolithic to Saxon, showed 74 per cent to be affected (Brothwell, 1959c). Frequencies are not, of course, always as high as this in earlier groups. Rusconi (1946), for

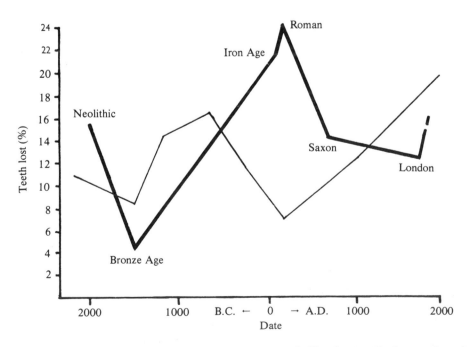

Figure 6.11   Tooth loss through disease in Britain and Greek populations of various Ages. British: ━; *Greek:* —.

example, noted that in 107 Indian skulls from
Mendoza, periodontal disease was present in
only 17.8 per cent. In a detailed study comparing Anglo-Saxon and seventeenth-century
British mandibles, Lavelle & Moore (1969)
noted significant increases in alveolar bone
resorption in the more recent series, a fact
they suggest is linked with a deterioration in
the nature of the diet.

Evidence of periodontal disease goes back
into Pleistocene times. A number of Neanderthal specimens show varying degrees of
alveolar resorption, an example from
Krapina being shown in Fig. 6.14. Slight
alveolar recession may also be present in the
Rhodesian skull, and indeed it seems
probable that this pathology has been
associated with man as long as caries.

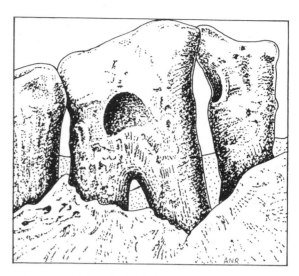

*Figure 6.13 Artificial non-carious inter-
proximal grooving.*

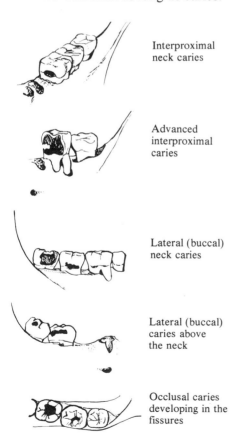

Interproximal
neck caries

Advanced
interproximal
caries

Lateral (buccal)
neck caries

Lateral (buccal)
caries above
the neck

Occlusal caries
developing in the
fissures

*Figure 6.12 Sites of caries development in
teeth.*

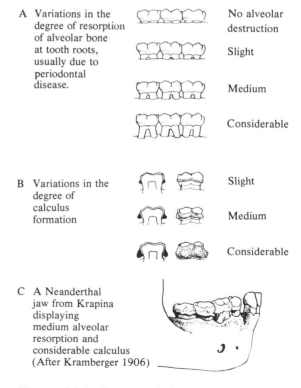

A Variations in the
degree of resorption
of alveolar bone
at tooth roots,
usually due to
periodontal
disease.

No alveolar
destruction

Slight

Medium

Considerable

B Variations in the
degree of
calculus
formation

Slight

Medium

Considerable

C A Neanderthal
jaw from Krapina
displaying
medium alveolar
resorption and
considerable calculus
(After Kramberger 1906)

*Figure 6.14 Stages of A, alvelor resorption
and B, calculus formation. C, mandible from
Krapina showing alveloar resorption and
calculus formation.*

iii) The chronic dental abscess

An abscess may be defined as a collection of pus, surrounded by denser tissue, and within a cavity of the body. Although oral abscesses may display a number of forms, the important factor as regards excavated material, is to detect any abscess cavities within the alveolus at the root apex, whatever their shape or size. The abscesses may form in association with general periodontal infection, considerable tooth wear or caries. It is not always easy to be sure of their presence, for small pockets at the tips of roots may pass unseen unless the tooth can be easily removed and the socket examined (which is not usually possible except with the incisors and canines). However, it is fortunate, at least for those who wish to examine early bone material, that abscesses do not usually remain inert, but the abscess 'burrows' out into the oral cavity, and in so doing, leaves clear evidence of its presence (Pl. 6.12). Often, in the region of the tip of a root, a clearly defined circular hole will be seen, extending into a larger cavity. This should not be confused with post-mortem erosion of the alveolus at the roots, which does not display such concise rounded margins (Fig. 6.15).

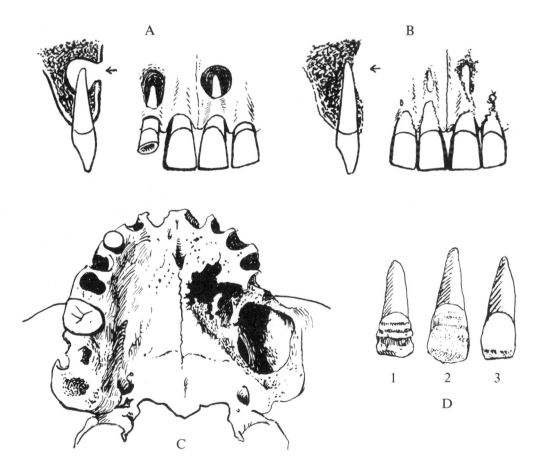

Figure 6.15   A, diagrammatic representation of chronic abscess cavities, showing the usual rounded cavity at the root apex, and opening to the exterior of the bone. B, post-mortem erosion at the roots which simulates abscesses. C, considerable destruction of a palate resulting from a large soft tumour. D, degrees of enamel hypoplasia: 1, considerable. 2, medium. 3, slight.

As with caries and periodontal disease, abscesses have been found in very early man. McCown & Keith (1939) found evidence of them at the roots of some of the teeth of Skhul V, from Mount Carmel. Rhodesian man had at least four abscess cavities, and possibly another two smaller ones. In the former specimen they were associated with periodontal disease, while in the latter, infection of pulp cavities by caries had probably resulted in the formation. Abscess frequencies vary considerably from people to people. Leigh (1925), for example, found that in excavated American Indian skulls 16 per cent of the Sioux specimens showed alveolar abscesses, whereas 52 per cent of the Zuni skulls had one or more. Similarly, Goldstein (1948) found in Indian crania from pre- and proto-historic sites in Texas that the West Texan cave or rock shelter peoples had

suffered far more from abscesses than the other groups. In contrast, British groups from Neolithic times onwards show remarkably constant and low frequencies.

Susceptibility to abscess formation also varies from tooth to tooth, and it seems possible that these frequencies also differ between groups. For example, various data collected by the author on early British material differ from the results obtained by Ruffer (1920) on Egyptians and suggests that Predynastic Egyptians were more liable to suffer from abscesses at the incisors but less so at the first molar than early British groups.

In the case of teeth with more than one root, abscess cavities may develop independently at each one. However, it is often very difficult to be sure of their independent origin, in which case only one cavity should be recorded.

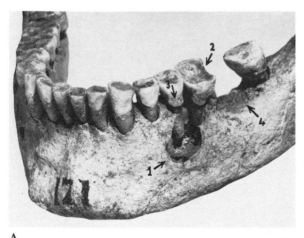

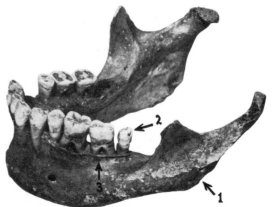

A   B

Plate 6.12   A, Bronze Age mandible from Yorkshire showing: 1, A large abscess cavity at the second premolar. 2, Typical 'hollowed-out' dentine wear. 3, Medium deposit of calculus. 4, Pre-mortem loss of the left second molar, with subsequent healing. B, A mandible from Dolgoi Island, Siberia, showing: 1, Noticeable hemiatrophy in the ramus region. 2, A very reduced third molar. 3, Considerable alveolar resorption through periodontal disease.

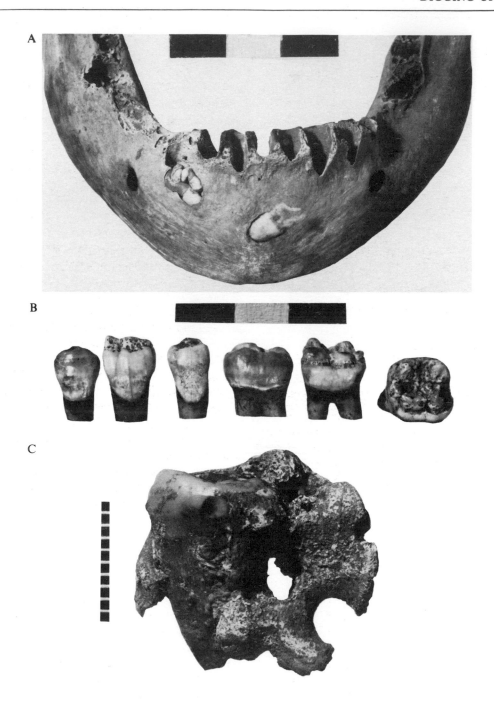

Plate 6.13 A, An odontome in the mandible of a female from Socotra. Note also the deflected tooth in mid-line. B, Teeth of an Iron Age skeleton from Ireland, showing varying degrees of enamel hypoplasia (also at different levels). The molar on the extreme right also has occlusal caries. C, An abnormally large calculus (tartar) deposit on a late Saxon molar from Nottinghamshire.

iv) Dental hypoplasia

Hypoplasia means underdevelopment, and when applied to the teeth is popularly thought to mean defective formation. Structural abnormalities may, however, be due either to causes acting before calcification of the tooth is complete, or to those occurring after. I propose to limit the term hypoplasia to the results of faulty structural development, which may be seen both microscopically and macroscopically. This does not include teeth abnormal in general shape or size, but only those which have deficient calcification, irregular distribution or partial absence of the enamel.

The degree of hypoplasia and number of teeth involved varies considerably depending upon:

a) Number of teeth already formed (these will not be affected).

b) Amount of crown formed before the upset occurred.

c) How long the casual factor is at work. This may be a short term disease such as scarlet fever, or deficiency in vitamin D.

Only the largest anomalies on the enamel surfaces are obvious to the naked eye, and these have been given the name of 'gross hypoplasia' (Mellanby, M., 1934). Usually this takes the form of 'bands' of depression or pitting on the tooth crown, parallel to the long axis of the mandible body. This is due to the fact that a tooth calcifies progressively from its occlusal biting surface to its root tip. It is possible to tell from the position and extent of such markings, as well as from the teeth involved, the approximate age of the individual when affected. As well as these transverse lines, cracks at right angles to the occlusal surface of the tooth are sometimes present. A good example of hypoplastic enamel is shown in Pl. 6.13, in teeth of an Irish Iron Age man from Faddan Beg, Co. Offaly. Not all cases, however, are as marked as this (Fig. 6.15), and it is therefore necess-

ary to look at the clean enamel surfaces carefully before one can be sure that no irregularity is present. By sectioning the teeth, one can examine the microscopic defects in both the enamel and dentine.

This defect is by no means restricted to modern civilized man, and in one Australopithecine genus, *Paranthropus*, 28 per cent of forty-seven isolated teeth showed signs of hypoplastic pitting (Robinson, 1952). Whether these defects denote childhood illnesses in one or two of these man-apes, or is evidence of seasonal malnutrition, remains uncertain. The teeth of Upper Pleistocene man were not always perfectly formed, and Soggnnaes (1956) found faulty microstructure in most tooth samples of the Mount Carmel people that he examined. Later groups show similar enamel defects, and in a series of early British skulls 58 per cent of those examined displayed some enamel defect in more than one tooth (Brothwell, 1959*c*). It is of considerable interest to note any degree of hypoplasia, for we are thus able to assess the health of earlier populations.

Although hypoplasia has mainly been studied in tooth crowns, defective formation can also occur along the root, and Taylor (1972) gives examples of constrictions and deformity in early Maori teeth.

v) Dental calculus (tartar)

This is a concretion that forms on the teeth, usually at the margin of the gums (Pl. 6.12). It consists mainly of a calcium deposit which, although fairly soft, nevertheless tends to persist on the teeth of earlier man. Care should be taken when cleaning specimens, that these deposits are not accidentally flaked off, for they easily break away from the smooth enamel surface. During life, the calculus may also contain food debris and bacteria, but this would seem to have little association with the production of caries. The deposits do, however, irritate the gums and help to initiate periodontal disease.

When present, the amount of the deposit should be noted roughly, that is, whether it is slight, medium or considerable (Fig. 6.14). If none is seen one should not conclude that the individual was without calculus, unless it is quite certain that the deposit has not been washed off, or eroded away in the soil. Any abnormal development of calculus should be noted separately; for example, when the deposit is abnormally large (Pl. 6.13), when the molar occlusal surfaces are also coated over (due to lack of use), or when the biting surfaces of the incisors are involved.

The fact that such deposits were common in earlier man is of considerable interest, especially as it may co-exist with severe attrition. Widespread calcareous coats no doubt bring some degree of immunity from caries to the person who had them, even though the gums would be irritated. If chewing surfaces, normally calculus-free, are coated this may be evidence of lack of occlusion of upper and lower teeth, or perhaps of a long-standing illness in which the person's oral health had severely declined.

vi) Cysts

Dental cysts are not, as at one time they were thought to be, purely chronic abscess cavities, but rather should be classed as a type of soft-tissue tumour. They are usually quite small and may be left unrecognized or be mistaken for simple abscesses. Owing to this difficulty in identification, any suspected cyst in excavated material should be referred to a dental specialist for further examination. In modern Europeans, they are more common in the upper than in the lower jaw.

Large cysts are more easily identified, and a number have been described from various early populations. In ancient Nubian material Elliot Smith & Wood Jones (1910) described a large cystic cavity in the alveolus of an edentulous female. Salama & Hilmy (1951) discussed two cysts in Fifth Dynasty material excavated at Sakara. The first case is a large dental cyst with palatal expansion, the second specimen a mandible with multiple cystic cavities. Risdon (1939) noted two cysts in the material from Lachish, and Goodman & Morant (1940) described two symmetrically disposed cyst cavities in the buccal surfaces of the maxillae of a Maiden Castle skull. An extremely large example of what may well have been a cyst was described by Moodie (1931) in a skull from Peru (Fig. 6.15).

vii) Odontomes

An odontome is a dental tumour, the microscopic structure of which may resemble the normal structure of a tooth (Keith, 1940). It may be embedded in the jaw or partially erupted, the latter condition being more liable to discovery in excavated material. Reviews of these tumours include Thoma (1946) and Hitchin & Mason (1958). Few such tooth-germ disturbances have been recognized in archaeological material. One definite case, now preserved in the British Museum (Natural History), is in a mandible from a cave on the island of Socotra (Pl. 6.13). The odontome is of the 'compound composite' type, showing a number of hard dental particles clustered together in a round cavity. One or two of the particles display both enamel and dentine and look very much like minute teeth (Brothwell, 1959b).

10   Skeletal deformities

i) Infantile paralysis (poliomyelitis)

The skeletal evidence of this disease is still uncertain, and of the symptoms so far put forward to substantiate its occurrence all may be equally well attributed to other causes. An abnormally short left femur in a Predynastic Egyptian (Sigerist, 1951), as well as foot deformities in a priest of the Eighteenth Dynasty and in the Pharaoh Siphtah, have also been given as evidence. Without discussing this in any more detail, it may be said that cases of shortening or thinning in bones from the lower limbs should not be immediately attributed to this disease — at least not until expert advice has been sought.

ii) Congenital dysplasia of the hip joint and sequelae

This group of disorders has been reviewed by Hart (1952). The term 'dysplasia' refers to abnormal growth in the hip joint, usually through extrinsic factors. The condition may or may not give rise to various secondary abnormalities including potential dislocation (subluxation) of the femoral head, dislocation (luxation), arthritic changes, coxa valga and coxa vara (angular anomalies of the femoral neck). As family studies have shown, dysplasia of the hip joint is in some cases genetically determined.

It seems probable that a number of hip abnormalities known in archaeological specimens may depend in the first instance upon a congenital dysplasia of the joint. Congenital luxation of the hip has been found in a Neolithic specimen from France, in an ancient Peruvian, and among the pre-Columbians of America (Pales, 1930). Elliot Smith & Wood Jones (1910) described five female specimens from Nubia that all displayed separated femoral-head epiphyses accompanied by dislocation of the rest of the femur. It seems unlikely that trauma alone caused these five anomalies, and a more reasonable explanation would be an inherited dysplasia. Angel (1946) also noted a single case of congenital hip dislocation in an Early Iron Age skeleton from Greece.

A British specimen is known that may also be due to a primary growth anomaly. This is a Saxon skeleton from Guildown (Pl. 6.10) showing quite clearly a flat inadequate acetabular socket, malformation of the femoral head (which does not 'fit' the socket well), and possibly some osteoarthritic change. The general impression given by the hip joint is that socket growth had been somewhat deficient (Brothwell, 1961b).

It is possible that some earlier groups may have been more prone to hip dislocation than others. Among the Lapps congenital dislocation of the hip occurs relatively frequently (Getz, 1955).

11   Abnormalities due to endocrine disorders

When certain glands of the body do not function properly, growth may be distorted in various ways. The resulting abnormalities depend upon whether the gland is producing too much of a certain hormone, or too little. The most important growth disorders are associated with the pituitary gland situated at the base of the brain, and may be classified according to the functioning of the gland.

i) Hyperpituitarism

This over-function of the gland, with excess of the growth hormone, has been usefully discussed by Weinmann & Sicher (1947) and by Fairbank (1951). It gives rise to two main types of abnormality, as follows:

a) Gigantism
Little need be said of this disorder, for the length of the body, or long-bones, should make the abnormality quite evident. Moreover, probably owing to its great rarity, no case has yet been found in earlier man. Massiveness of bone, as seen in the Pleistocene hominid fossils *Meganthropus*

and *Paranthropus robustus* is not of course evidence of disturbed glandular function, but of genetically determined size-adaptations.

b) Acromegaly
Whereas gigantism develops during childhood and adolescence, acromegaly is mainly a disease of adults. A tumour of the pituitary gland reinitiates growth, especially in the skull, hands and feet; it also results in enlarged vertebral bodies and ribs, and in considerable thickening of the long-bones.

A   B   C

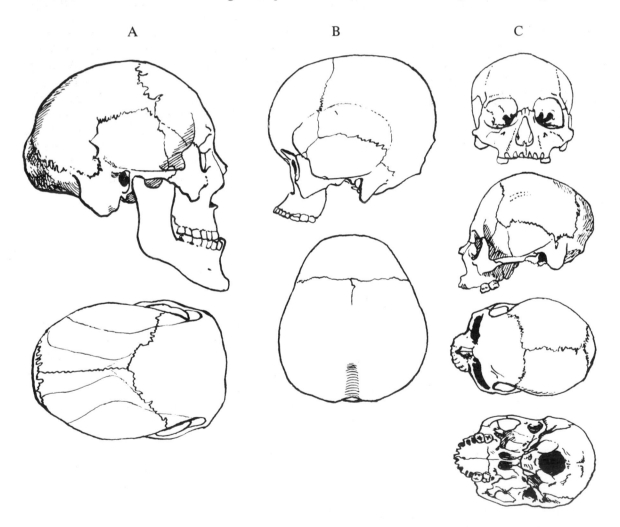

Figure 6.16   A, acromegalic skulls. Lateral view of an acromegalic giant (Cornelius Magrath). Top view of an acromegalic skull showing the very high temporal lines. B, lateral and top views of an early Egyptian achondroplastic skull. C, four aspects of the skull of a midget from Peru.

The skull displays the most characteristic features, which include the enormous development of supraorbital ridges and frontal sinuses, the lengthening of the face due to additional growth in the mandibular and palatal regions; while the occipital bone in the region of the nuchal crests becomes greatly thickened (Fig. 6.16).

A case of acromegaly in a Neolithic individual from Switzerland has been described by Schlaginhaufen (1925), but the diagnosis has since been contested. A more certain case, probably early Egyptian, is preserved in the British Museum (Natural History). The face is extremely lengthened (Pl. 5.1) although the supraorbital ridges are not abnormally large, perhaps because the person was female. Keith (1931) has described a twelfth-century skull from Gardar, Greenland, which from the evidence of vault and mandibular fragments, appears to be another example of this disorder. Cases of early fossil man have also been likened to acromegalics, but although the growth processes producing certain bony forms may have been similar, it would of course be absurd to consider that they all suffered from disturbed pituitary function.

ii) Hypopituitarism

This disorder, sometimes known as pituitary dwarfism, is due to deficiency of the growth hormones. Most midgets, other than achondroplastics, have small but normally proportioned bodies. They may be easily identified by their size and slender bones. Skulls of midgets, mainly Peruvian examples, have been discussed in detail by Hrdlička (1939), and one case is shown in Figure 6.16. Small crania of the Third Dynasty from Egypt have also been decribed (Smith, 1912), but these may in fact have been pygmy individuals and not persons suffering from hormone deficiency. If only the skull vault is present it may be difficult to decide whether the individual was a dwarf or merely microcephalic.

12   Effect of diet on bone

Bone growth and size may be influenced either by a completely inadequate diet or by deficiency of the 'calcifying' vitamin D. In the former case, starvation of the body over a period of time may result in slenderness of bone and smallness of stature, even excluding the results of a specific vitamin deficiency. More restricted diet, in regard to variation but not quantity, may lead to the development of severe disorders.

i) Rickets

During infancy the bones are 'soft' and 'pliable', becoming gradually hardened with age, mainly by the deposition of calcium phosphate. For calcification to proceed normally, the body must be given sufficient vitamin D (Mellanby, E., 1934), this substance being particularly concentrated in fish liver oil, as well as being produced in the superficial layers of the skin by the ultraviolet rays of the sun. If there is a calcium deficiency, then of course, no amount of vitamin D can ensure good bone.

Rickets became particularly common in industrial parts of Europe during the eighteenth and nineteenth centuries, when many working-class families were compelled to live on inferior diets, while the children did not enjoy sufficient sunshine to obtain the needed vitamin D. The result was that many children suffered from this unpleasant disease. However, this must not be considered purely a disease of recent times, although evidence of it in earlier man suggests that it was far from common, especially in warmer climates where sunshine amply made up for deficiencies of this vitamin in food.

Perhaps some of the earliest evidence of rickets is to be found in ancient Egyptian wall illustrations, where bow-legged individuals are represented. Elliot Smith & Wood Jones (1910) pointed out that although no definite cases of rickets have been found in ancient Nubian cemeteries, many of the bones examined 'exhibit distortions difficult to explain except by invoking rickets as the causal agent'. Indications of the disease have also been found in Neolithic bones from Denmark, Germany, and Norway (Sigerist, 1951; Grimm, 1972), and Rolleston described human remains that he considered to show signs of rickets (Wright, 1903). Whether all these cases are truly diagnosed is debatable, but it seems reasonable to suppose that with the development of towns and cities in Europe some severe cases of malnutrition occurred periodically, especially during the winter months. Certainly when examining early skeletal material, especially of the last few centuries, evidence of vitamin deficiency must be looked for.

Rickets may be present in varying degrees, the following being a few of the characteristic features (Dick, 1922):

a) General retardation of skeletal growth.

b) Development of frontal and parietal 'bossing'.

c) The bone is light and brittle in texture.

d) The arch of the palate may be abnormally high.

e) The femur is, as a rule, curved forwards and outwards throughout its length, an exaggeration of the slight normal curve of this bone (Fig. 6.5). The tibia, fibula, ulna and radius may also be curved (not to be confused with post-mortem deformation).

f) Knock-knee or bow-legs may develop.

g) Asymmetry and distortion of the chest (pigeon-chest) is a common defect, this rib distortion usually being the result of a slight lateral curvature of the spine (scoliosis).

h) Scoliosis (Fig. 4.18) is the lateral deviation of the spinal column, the curve being either to the right or to the left.

ii) Osteomalacia

This disease of adults is similar to rickets, in that there is lack of ability to calcify bone. Causes other than a simple lack of vitamin D are known (Jackson, Dowdle & Linder, 1958). The bones may be thin and are noticeably light owing to demineralization. The pliability of bone may result in platy-basia (the base of the skull bends upwards), pigeon-chest, knock-knees and scoliosis.

Few cases have yet been reported in excavated skeletons, although its prevalence in areas such as India at the present time make it likely that cases will be found. Two early examples have so far been found in Peru (Ackerknecht, 1953).

iii) Other diseases possibly associated with diet

Although dietary variations affect the body in many ways, few others change the bones to any marked degree. It has been suggested by Ackerknecht (1953) that osteoporosis (pitting) of the skull, discussed elsewhere (p. 166), might be due to avitaminosis.

Gout (see p. 151), a condition noted in a Coptic mummy by Elliot Smith and Dawson (1924), has now been identified in a Roman skeleton from Cirencester. Changes to various parts of the limb skeleton include joint destruction and small cavitations in the long-bone cortex (Wells, 1973).

13   Bone changes associated with blood disorders

A number of anaemic diseases leave their mark on bones. Two of these are particularly important in that they are hereditary abnormalities having a noticeable frequency in two broad areas of the world at the present time. Studies of ancient bones in these and other areas of the world thus provide an opportunity to continue the study of two distinctive anthropological traits backwards in time. The task is not an easy one, for the changes taking place in the bones of the affected individuals resemble changes resulting from other causes. Fairly confident differentiation of these two anaemias from other abnormalities nevertheless seems a possibility, and Neel (1950) considers that investigations along these lines are worth while.

The two blood disorders referred to are *thalassaemia*, also called Mediterranean anaemia because it is so common in this part of the world, and *sickle cell anaemia*, which is mainly restricted to Negro groups. Although a recent study of bone changes in the latter disease shows that various post-cranial changes may take place (Carroll, 1957), nevertheless the most apparent changes in both are in the skull vault. The individuals show a marked increase in the thickness of the vault, and radiographs of the thick bone often show perpendicular striations sometimes called the 'hair-on-end' effect. Two Nepalese skulls in the British Museum (Natural History) (Morant, 1924) could be examples of such anaemic reactions (Fig. 6.6), for it seems unlikely that rickets could have affected these adults in this way. Sir Arthur Keith also examined these specimens and concluded that this was not a case of the rare condition *leontiasis ossea*. Anaemic reactions have been reviewed in connection with the Piltdown skull and Oakley (*in* Weiner *et al.*, 1955) suggests that the thickening of the diploë in this cranium may also be 'a reflection of a severe chronic anaemia'. Since then, there has been growing interest and work on this condition, usually known as *porotic hyperostosis* (Angel, 1966), and indeed of the whole question of skeletal osteoporosis of the kind not related to senility (Angel, 1964).

Cribra orbitalia *(Orbital osteoporosis)* (Fig. 6.17) has recently been discussed in some detail by Møller-Christensen (1953), Nathan & Haas (1966), Knip (1971), and Hengen (1971). These strainer-like perforations in the roof of the orbit (Pl. 6.6) have been noted in various series of skulls, the earliest extensive work being by Welker (1887). The exact cause of this anomaly is not known, although it has been suggested that it is due to the pressure of an enlarged lacrimal gland. The fact that it has been found to a minor degree in the newly born suggests that it need not be caused by a common deficiency disease, although this does not rule out an environmental origin. However, as the frequencies are quite variable the trait seems well worth studying, and even if it is shown to be caused purely by the external environment, it may still be of value as an environmental 'indicator'. It would now seem advisable always to consider it separately from non-metrical data, and rather to include it in relation to other cranial pathology.

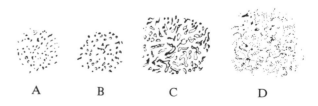

A   B   C   D

Figure 6.17   Schematic representation of the three main types (degrees of development) of the cribra orbitalia. A, porotic type: scattered, isolated fine apertures. B, cribrotic type: conglomerate of larger but still isolated apertures. C, trabecular type: apertures confluent resulting in the formation of bone trabeculae. D, closed-trabecular type: apertures are closed, but the surface is crossed by sulci and depressions (after Knip, 1971).

14   Bone affections of uncertain origin

i) Paget's disease

This chronic disease, sometimes known as osteitis deformans, may develop in one or more bones, usually after the age of fifty years. The most common sites for the disease are the pelvis, femur, tibia, lower spine and skull (Fairbank, 1951). Although radiographically various forms of bone change may be noted, the principal change is some degree of thickening, with the denser cortical (outer) bone being replaced by spongy bone. The long-bones may become bowed, and if the skull is affected, considerable thickening is common (Fig. 6.5). Diagnosis may not be easy unless a number of bones are involved.

As yet only one prehistoric case of this disease appears probable. This is in a French Neolithic femur from Lozère (Fig. 6.5). The bone, especially the shaft, is modified in form and structure. The shaft is slightly bowed and noticeably swollen; much of the outer compact bone has been replaced by a honey-combed network (Pales, 1930). It is not surprising that this is the only recorded case dating from prehistoric times, for if we assume that the time of onset was the same then as now, a large percentage of the population would never have reached the required age. From more recent times, two possible cases have been discovered in the excavations of medieval Winchester.

ii) Osteoporosis

As the name implies, the condition involves abnormally porous bone, either in a restricted area or widespread throughout the skeleton. Various kinds of so-called osteoporosis are in fact known. Senile osteoporosis, for example, is a generalized process of demineralization of the bony framework of the body, usually developing after the sixth decade. Age-related bone loss is now known to occur to a lesser extent from the fourth decade of life onwards, with differences appearing related to sex as well as age. (Garn, 1973; van Gerven, 1973).

This question of changes in cortical bone has now been fruitfully studied in relation to archaeological samples from Africa and the New World (Dewey, Armelagos & Bartley, 1969; Carlson, Armelagos & van Gerven, 1976). However, the anthropologist is mainly concerned with the detection of 'osteoporitic pitting' that is found on certain bones. In a study of American War Dead from Korea (McKern & Stewart, 1957), this form of osteoporosis was found on the skull vault, interproximal surfaces of the vertebrae, margins of the sternum, and on the medial surface of the clavicle; but rarely on all these bones in the same individual.

By far the commonest area to show osteoporitic pitting is the skull (Pl. 6.6), where it is usually bilateral and symmetrical. The parietals are particularly susceptible, less frequently the frontal and occipital, and rarely the facial and sphenoid bones. A puzzling form of this symmetrical osteoporosis, namely cribra orbitalia (Pl. 6.6), has just been discussed. It is possible that children are more easily affected, and it is suggested by Cybulski (1977) that the high incidence of 53 per cent in immature Haida and other Canadian samples is suggestive of an iron-deficiency anaemia. This was the conclusion of Carlson, Armelagos & van Gerven (1974) working an ancient Nubian material.

Cases of osteoporitic pitting are known in Europe from Neolithic and later periods. Pre-Columbian American Indians were especially susceptible (Pales, 1930). The reason for its development is not understood, but it is probable that the answer will be complex. As the condition is clearly associated in some cases with environment, it may yield information about the health of earlier man.

It is not known whether parietal thinning (Brothwell, 1967b; Satinoff, 1972) may be linked to parietal boss osteoporosis – perhaps as a sequel? The bilaterally symmetrical depressions may result in the thinning of the parietals to a millimetre or less.

15   Congenital defects or abnormalities

i) Achondroplasia

Achondroplasia or chondrodystrophia foetalis is the name given to the commonest form of dwarfism, and is a congenital condition that results in abnormal bone growth. The individual is usually under 1.2 metre in height, even when fully adult, the shortness being mainly due to the small length of the lower limbs. The long-bones of an achondroplastic may be easily distinguished from other forms of dwarfed bone by their noticeable thickness relative to length (Fig. 6.18). The head is as large, if not larger than in normal individuals, and the frontal region is prominent. The face may be relatively small, and the bridge of the nose depressed and flattened. Other abnormalities possible include short and broad hands, bowed legs, spinal curve defects and small flat chests.

Dwarfs of this kind are known to have occurred from earliest times in Egypt. Jones

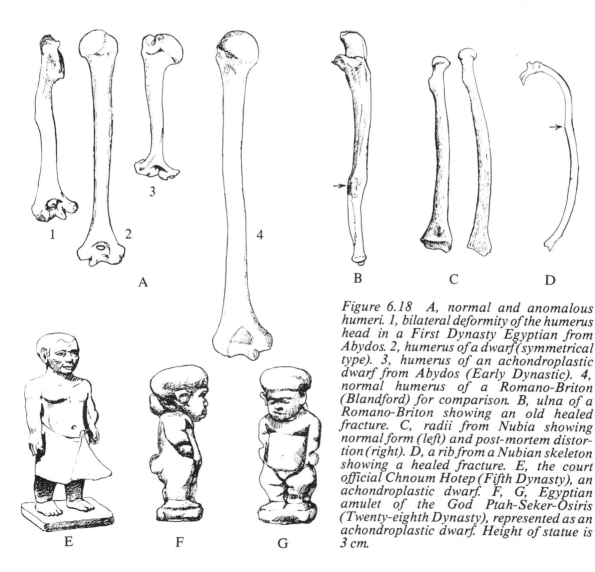

Figure 6.18   A, normal and anomalous humeri. 1, bilateral deformity of the humerus head in a First Dynasty Egyptian from Abydos. 2, humerus of a dwarf (symmetrical type). 3, humerus of an achondroplastic dwarf from Abydos (Early Dynastic). 4, normal humerus of a Romano-Briton (Blandford) for comparison. B, ulna of a Romano-Briton showing an old healed fracture. C, radii from Nubia showing normal form (left) and post-mortem distortion (right). D, a rib from a Nubian skeleton showing a healed fracture. E, the court official Chnoum Hotep (Fifth Dynasty), an achondroplastic dwarf. F, G, Egyptian amulet of the God Ptah-Seker-Osiris (Twenty-eighth Dynasty), represented as an achondroplastic dwarf. Height of statue is 3 cm.

(1931) described a case from the Badarian period, showing the characteristic shortening of the long-bones, although there was no noticeable change in the skull. A skull from the Eighteenth Dynasty (Fig. 6.16) shows the typical depressed nasal region (Keith, 1913a), although it was originally diagnosed as a cretin (Seligman, 1912). The court official Chnoum Hotep, who lived at the time of the Fifth Dynasty, is clearly an achondroplastic as his limestone statuette shows (Fig. 6.18). Other cases in excavated material are known, and it is not to be thought that Egypt is the only region where the anomaly occurs, although as yet no early British examples are known.

ii) Hydrocephaly

If a skull is noticeably larger than usual, the individual may well have been afflicted with some degree of hydrocephaly (commonly known as 'water on the brain'). The condition need not always be of congenital origin. The vault, but not the face, is uniformly increased in all directions, assuming a very globular shape. The sutures may be widely separated or, in arrested cases of the disease, wormian bones may help to fill the sutural gaps and unite the adjoining bones. Care must be taken in differentiating between increase in vault size due to rickets and anaemia, which produce skull thickening, and hydrocephaly where there is more often thinning of the cranial bones. Various causes are known for hydrocephaly, but need not be discussed here. In a study of modern British cases, it was found that the disease usually becomes evident in the first six months of life, and the death rate is highest during the first 18 months (Laurence, 1958).

Slight cases of possible hydrocephaly, as found in First Dynasty material from Sakkara (Batrawi & Morant, 1947) and Iron Age Lachish (Risdon, 1939), are difficult to confirm. Three cases, however, are indisputably due to this disease. A British example has been described by Trevor

(1950b). The skull is of a Romano-British male from Norton, Yorkshire (Pl. 6.14), having a maximum length of 216.1 mm, a maximum breadth of 176.3 mm and a basio-bregmatic height of 178.7 mm. The cranial capacity was probably in the region of 2600 cm^3, which is over 1000 cm^3 more than the average for Romano-British man. Derry (1913) has described a case of hydrocephaly associated with post-cranial anomalies in an Egyptian of the Roman period, and Grimm & Plathner (1952) described a case from Germany in a child from the Neolithic period. As the intelligence of modern individuals affected with this disease varies considerably from normal to ineducable, it cannot be said that these archaeological specimens represent 'village idiots' of the past.

iii) Acrocephaly

Perhaps the most constant deformity of this abnormal skull type is the very high and brachycephalic vault (Fig. 6.19). Acrocephaly, sometimes called Oxycephaly (meaning tower-skull) has been very extensively discussed by Greig (1926) who considers that the condition may be congenital or the result of injury and disease. There is no doubt that here, as in scaphocephaly, the post-natal environment may profoundly disturb vault growth, but doubtless many cases are of a congenital nature. The disorder is mainly dictated by abnormal growth at the coronal suture, but various other sutures may be involved to a greater or lesser degree. Instead of allowing the normal amount of longitudinal (antero-posterior) growth at the coronal suture, there is a premature fusion. This prevents the vault from attaining its natural length, and there is instead compensatory growth upwards. This is sometimes associated with considerable deformity of the hands and feet (acrocephalo-syndactyly), but this latter form is far more rare and has not been discovered in early man.

Many specimens of acrocephaly are known, their vault form being so easy to distinguish. Keith (1913a) figures a well-marked case in an Early Dynastic Egyptian. Hug (1956) also figures an early case from Bern, but in this specimen there appears to be an associated depression in the region of lambda (known as bathrocephaly, Fig. 6.19), which is often found as an independent anomaly, especially in the seventeenth-century London skulls (Macdonell, 1906; Hooke, 1926).

iv) Microcephaly

The term microcephaly, or small headedness, is perhaps one of the vaguest terms in medicine. It can be used to refer to any skull that is noticeably below size, say with a cranial capacity of less than 1000 cm³. Although there is a well-defined condition called 'true microcephaly', the word is retained here to refer to any small skull. This, then, will include mentally defective individuals such as mongolian idiots as well as some intelligent midgets. When possible, the exact nature of the microcephaly (i.e. whether idiot or dwarf) should be stated.

Although many microcephalics of the idiot type must certainly have been killed or died of neglect in earlier times, a few cases are known. One of the earliest cases is that of an aeneolithic individual from Sano Cave in Japan (Suzuki, 1975). This was an adult

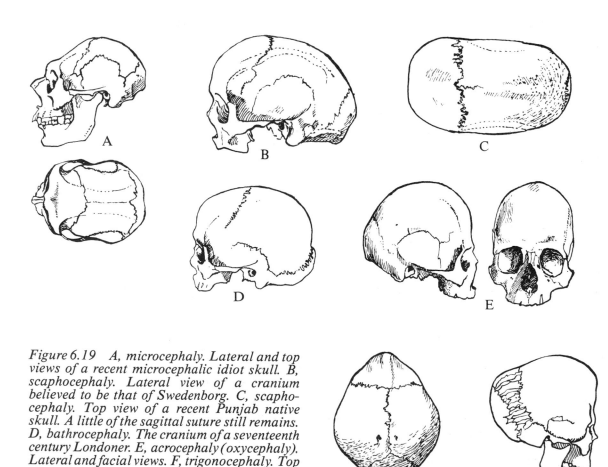

Figure 6.19   A, microcephaly. Lateral and top views of a recent microcephalic idiot skull. B, scaphocephaly. Lateral view of a cranium believed to be that of Swedenborg. C, scaphocephaly. Top view of a recent Punjab native skull. A little of the sagittal suture still remains. D, bathrocephaly. The cranium of a seventeenth century Londoner. E, acrocephaly (oxycephaly). Lateral and facial views. F, trigonocephaly. Top view of a modern example. G, lateral view of an acrocephalic early Egyptian skull.

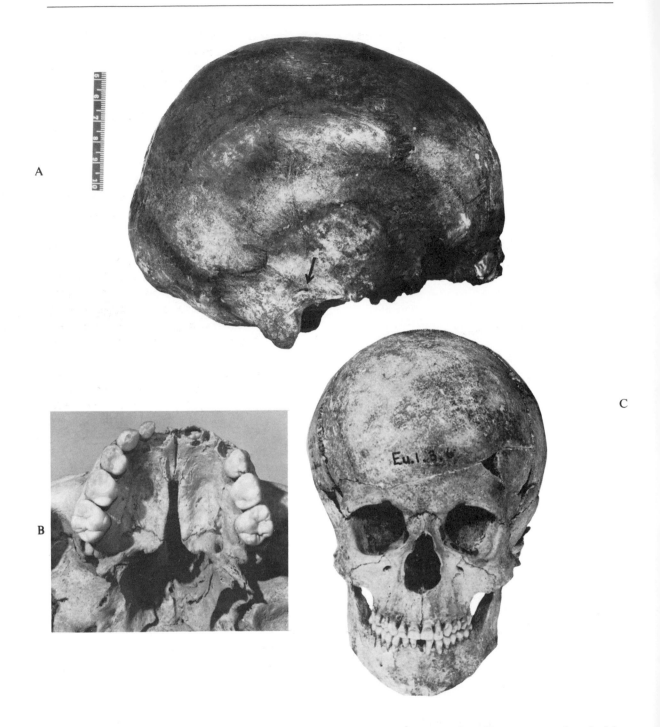

Plate 6.14   Congenital anomalies. A, Complete absence of the external auditory meatus (ear hole) in an Iron Age skull from Lachish, Palestine. B, Cleft palate in a Saxon child from Burwell, Cambridgeshire (by courtesy of the Duckworty Laboratory, Cambridge). C, A Romano-Briton from Norton, Yorkshire, showing noticeable hydrocephaly (by courtesy of the Duckworth Laboratory, Cambridge.

individual with a brain size of about 730 cm³. Elliot Smith & Wood Jones (1910) noted a 'microcephalic' female who had been excavated from a burial pit of the New Empire (Eighteenth–twentieth Dynasties) of Egypt. One of the individuals in the mass burials found at Donnybrook, near Dublin was described by Frazer (1892) as microcephalic, the burials probably having taken place during the tenth century (Martin, 1935). Yet another case is to be seen in the medieval ossuary at a church in Hythe, and may be as early as twelfth century A.D. Finally, a specific form of microcephaly denoting mongolian idiocy has been found in a late Saxon cemetery in Leicestershire (Brothwell, 1960b).

v) Other congenital anomalies

Much less common abnormalities include osteopetrosis (so-called 'marble bone disease' because of the excessive bone density); diaphysial aclasia (multiple exostoses associated with abnormal bone forms); cranio-cleido-dysostosis (deficient clavicle and skull formation, sometimes with other defects). Their occurrence in earlier populations is as yet poorly known.

A number of minor deformities have, however, been noted, some of which may be listed briefly as follows:

a) Absence of a whole bone or part of one. The nasal bones are particularly apt to be missing or deformed (Fig. 2.25) (see Wahby, 1903). Absence of the basi-occipital has been described in a Romano-Briton (Brothwell 1958a), while a cleft palate has been noted in a Saxon skull from Burwell, Cambridgeshire (Pl. 6.14), and Elliot Smith & Dawson (1924) also record a case in an ancient Egyptian. Incompletely-formed vertebrae, especially in the neural arch region (Fig. 4.18), have been found in skeletons from various parts of the world.

b) Congenital fusion of bones may take place at various parts of the skeleton. Cases of rib fusion (Fig. 6.20) and atlas vertebrae

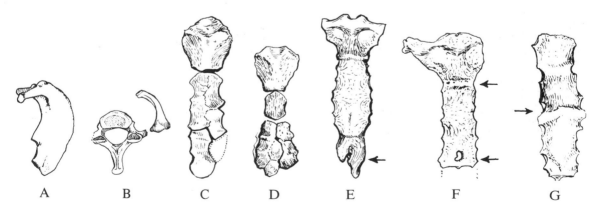

Figure 6.20   A, anomalous ribs (fused except in the articular region) from an Australian aboriginal. B, seventh cervical vertebra and its cervical rib from a Romano-Briton. C, D, anomalous sterna from early Nubian skeletons. In each case, the sternal body is composed of a number of segments (after G. Elliot Smith & F. Wood Jones). E, ossified xiphosternum. Fairly common. F, sternum of an Anglo-Saxon, showing a perforation of the sternal body, and a fused manubrium. G, sternal body of an early Nubian, showing an old healed fracture (after G. Elliot Smith & F. Wood Jones).

joined to the occipital bone were found in early Nubian material (Elliot Smith & Wood Jones, 1910).

c) Extra bones, excluding wormian bones, are not common. In modern populations presacral vertebrae usually number 24, but 25 can occur, the frequency ranging between 2 and 12 per cent in different groups (Allbrook, 1955). Of less frequent occurrence is the cervical rib (Fig. 6.20). Six cases of this rib anomaly were found in early Nubian material, and Denninger (1931) mentions other cases as well as describing in detail a specimen from a prehistoric Indian mound in Illinois. Angel (1946) notes a left cervical rib in an ancient Greek skeleton. Occasionally, the body of a sternum is found to be segmented (Fig. 6.20) in the adult.

d) Multiple exostoses, possibly the result of a genetic factor, have been found in a pregnant woman of medieval date from Sweden (Sjøvold, Swedborg & Diener, 1974).

16   Abnormal skull form

The synostosis (fusion) of cranial sutures may be either normal, precocious or retarded. If a suture ceases to grow and ossifies prematurely, deformation of the skull form results and the degree of distortion is to some extent dependent upon the age of the child when fusion took place. In the less extreme cases the rest of the skull usually compensates the localized restriction in growth, although slight microcephaly (small-headedness) may result. This anomaly may be present at birth (congenital) or it can be initiated by injury or disease.

i) Scaphocephaly

This is by far the most common of the synostotic deformities, examples being known from Neolithic times onwards. On examination, the skull vault is found to be abnormally long and narrow (the cephalic index usually being well below 70). This is due to lack of growth along the sagittal suture which is either completely or partially obliterated (Fig. 6.19). Some degree of medial ridging (keeling) is also often found in these skulls, the parietal bones being more flattened than usual. Any signs of injury in the sagittal region should be noted, and vault measurements should be excluded in metrical comparisons. One of the skulls claimed to be that of Swedenborg is a good example of scaphocephaly (Fig. 6.19).

ii) Other abnormality

The metopic suture is normally present at birth and is obliterated within the first two or three years. If this suture undergoes very early closure (even before birth), the frontal bone does not attain its normal breadth but is very narrow above the orbits and expands towards the coronal suture (Fig. 6.19). This anomaly, called trigonocephaly, is far more rare than scaphocephaly.

If the vault of the skull is very noticeably asymmetrical, it is said to be plagiocephalic. Duckworth (1915) states that in a typical case there is unilateral and premature arrest of growth along part of the coronal suture, but there are probably other causes, such as variations in the growth rates of the cranial bones. The result of this defective growth is an asymmetrical vault. The frequency of this anomaly is probably far less than some earlier authors imply, and it is more usual to find that gross asymmetry is the result of artificial (ante- or post-mortem) deformation. Considerable changes in cranial form are possible even when the bones are not decalcified (Fig. 6.21).

17   Giants, dwarfs and illusions

As in other archaeological fields, interpretation of skeletal remains should only follow after a factual analysis. It is only too easy to develop exciting theories, particularly if they appear to fit in with the cultural picture one has in mind. It is usually said, for example, that the sizes of armour in the Tower of London show that in Medieval times the average stature was much smaller than it is today, although in fact skeletal evidence does not seem to substantiate this claim. Similarly, the Anglo-Saxons are commonly throught to be tall people and we sometimes find that the overall height of a skeleton within a grave (from the top of the head to the tips of the toes) appears to be well over 1.8 metres. However, it is often found on examining the long-bones in the laboratory that the individual was in fact a number of centimeters shorter; the misconception has resulted from the fact that with decomposition of the soft tissues the bones had spread, giving a false impression of body size. Also, it is sometimes tempting to suggest an abnormally large individual on the slender evidence of a particularly robust skull, mandible or post-cranial fragment. The Australo-pithecines of the genus *Paranthropus* had excessively massive jaws, yet they were of relatively small stature; similarly *Gigantopithecus* distinguished by enormous teeth, was not necessarily gigantic in stature.

Dwarfism is another condition that may quickly conjure up magico-religious or ritual associations, and there is no doubt that these little people have held special positions in some courts of the past. The ancient Egyptian god Ptah was sometimes represented in the form of an achondroplastic dwarf. With incomplete skeletal material it is easy to mistake immature remains for evidence of dwarfism. An example of this is to be found in the crouched skeleton uncovered at the bottom of the outer ditch at Windmill Hill. This was for some time believed to be a dwarf, but on closer examination of the skeleton, both teeth and post-cranial remains indicated this individual had not been more than about three years old, with height well within the range for normal modern children of this age.

More localized effects upon bones may also give rise to unsound hypotheses. These may be discussed separately as follows:

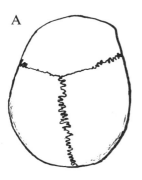 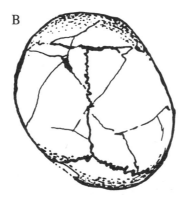 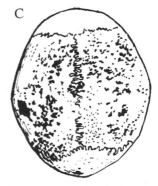

Figure 6.21   A, typical case of plagiocephaly. B, considerable distortion in a Bronze Age skull vault ('pseudo-plagiocephaly') due to earth pressure. C, pitting of the vault in an early Nubian, simulating osteoporosis, but caused by insects (after G. Elliot Smith & F. Wood Jones).

1. As the body decomposes (within or without a coffin) the femur head, humerus head and some other long-bone articular areas, may fall out of position, suggesting a severe dislocation during life. This may be checked or disproved to some extent by examining and articulating the bones; for unless the deformity resulted just before death, bone-changes may be expected to have occurred in the area involved.

2. Decomposition may also affect the appearance of the vertebral column, sometimes giving the impression of pathological deformity. Here again closer examination should confirm or disprove first impressions.

3. Severe crushing of the cranial vault may result in an extremely distorted form. Thus in certain aspects it may make it look longer, shorter or even larger than it really is. Such distortion can usually be distinguished by the over-lapping, lack of contact or gross asymmetry of the assembled fragments.

4. Variations in soil and climate have a variety of effects on bone. Pits formed by post-mortem corrosion of the surface of the vault have more than once been mistaken for evidence of syphilis, and occasionally a large and clearly-worn hole formed by a natural eroding agency may look very like a trephined area. Areas of teeth may also have been corroded in soil and may be mistaken for caries or some unusual disease of the enamel and dentine.

5. Cannibalism is particularly attractive to the imagination, but although occasionally evidence of this in European archaeological material has been claimed, there is little real proof of it (Brothwell, 1961c). For example, the absence of the base of the cranium has been quoted as evidence of brain-eating, but in fact this region is most easily broken and liable to decomposition. Similarly, the association of skeletal fragments with a hearth or rubbish pit cannot be taken as evidence of cannibalism merely because of their situation, although this could be the case if the fragments were cut and scratched. Again, isolated bones or fragments do not imply that the individual had been dismembered and his flesh eaten. As Cunnington (1949) pointed out, scattered human fragments may possibly result from the disturbance of burials that had not been clearly marked on the surface and whose exact location had been forgotten. Nor can we say that certain groups were cannibalistic merely because they concentrated a series of skulls, as in the Mesolithic 'nest' of skulls at Ofnet, Bavaria, or because the vault of a skull was given special burial, as appears to have been so in the Upper Palaeolithic rock shelter at Whaley, Derbyshire (Brothwell 1961a). Lastly, isolated skulls need not suggest the beheading of the victims; it is usual in such cases to find one or two cervical vertebrae still articulated with the occipital condyles, and to see the marks of an implement.

VII Concluding remarks

A sympathetic word to the archaeologist

What has been said in the preceding pages may seem at first a formidable study for the non-specialist. In fact the general information needed at any one time is not great, and this handbook is designed to be both a work of reference and a practical guide to the excavation and treatment of skeletal remains. Ideally the excavator should be fully acquainted with means of cleaning and preserving specimens, and with how to obtain the measurements and recordings that may help in their interpretation. Also, if the bones of the skeleton have been learnt, gross abnormalities should be easy to detect. Whether the anomaly is congenital or due to postnatal disorders is usually a matter for the specialist to decide; so too is the attribution of an osteitis to a specific disease.

Printed blanks are invaluable in osteological work, and Table 8 is given as an example of a sheet for recording primary data, once back in the laboratory.

Attention must first be drawn to the need for entering full details of the archaeological period and associated finds (or publications). It is often painfully clear, at least when examining early British material and the related records, that the excavator, or museum curator, has not thought of recording such details. Yet at the time when the finds were made they must have been well known. Excavators come and excavators go, but the human finds tend to remain for many more years, especially the badly labelled ones!

When preservatives have been applied, a note should be made of the precise kind that was used. There should be a record of whether any bone or soil samples have been taken for analysis. It is preferable that the samples should be sealed in air-tight tins or polythene bags, and not placed in matchboxes or envelopes.

If parts of the skull show post-mortem distortion, measurements involving these areas should not be taken, and if there is any uncertainty, a question mark should be placed against the reading. Even if it is not planned to store the information by computer, it is important to establish as much information as possible in a precise (and if needs be, coded) form. Thus, it is suggested that for those without 'caliper-to-computer' facilities, the data should be boxed-in as much as possible (Table 8).

To be sure of correct dental readings it is essential to have the jaws and teeth as clean as possible, but avoid removing calculus deposits. If there is any doubt as to the presence or absence of mild periodontal disease or a small caries cavity, the local dentist may usually be relied upon to be of help. The pattern of attrition may be checked against the range given in Figure 3.9 and the appropriate number booked down, or the actual attrition 'pattern' can be sketched.

Some of the non-metrical features can be quickly recorded on the form by '0' or '1' (0/1; 1/1 for both sides), depending on whether present or absent. In the case of extra bones along the lambdoid suture, ordinary wormian bones should be distinguished from an Inca bone. It is as well to make a small drawing of any epipteric bone present, for there may be a variety of shapes and positions of this bone at pterion.

The formula used in stature estimates will depend upon the population affinities of the group or individual studied. It is important to make a note of the formula used, otherwise future investigators will not know to what extent the method used was up to date.

Evidence of injury or disease should be described as fully as possible, and in the latter case it is more important to make quite clear the nature of the change rather than try to

Table 8

An example of a recording sheet for basic use with skeletons

PERIOD	ARCHAEOLOGICAL REFS	FIELD OR MUSEUM NUMBER
LOCALITY	PRESERVATION	DEFORMATION (a.m. or p.m.)
SEX	AGE AT DEATH	STATURE (+ ref.)

DENTITION

	ATTRITION
8 7 6 5 4 3 2 1 | 1 2 3 4 5 6 7 8	M_1 $\qquad$ M_2 $\qquad$ M_3
8 7 6 5 4 3 2 1 | 1 2 3 4 5 6 7 8	
χ = loss a.m. C = caries Ʊ = unerupted	U $\quad$ L $\quad$ U $\quad$ L $\quad$ U $\quad$ L
/ = loss p.m. A = abscess Ó = erupting	
E = pulp exposure	

CARIES in	ABSCESSES in	ENAMEL HYPOPLASIA in
TOOTH LOSS a.m. in	PERIODONTAL DISEASE	CALCULUS in
WORMIAN BONES	INCA BONE	METOPISM
PARIETAL NOTCH BONES	EPIPTERIC BONES	PTERION
TORI MANDIBULARES	TORI AUDITIVI	TORUS PALATINUS
TORI MAXILLARES	OSTEOPOROSIS (ORBITS)	INJURY
DISEASE (*a*) Arthritis		
(*b*) Other changes		

BASIC MEASUREMENTS

			Orbital B	(O'_1)
SKULL			Orbital Ht.	(O_2)
Maximum L	(L)		Palatal L	(G'_2)
Maximum B	(B)		Palatal B	(G_2)
Basi-Breg. Ht.	(H')		FEMUR	
Frontal Arc	(S_1)		Maximum L	(FeL_1)
Parietal Arc	(S_2)		TIBIA	
Occipital Arc	(S_3)		Maximum L	(TiL_1)
Basi-Nasal L	(LB)		HUMERUS	
Basi-Alv. L	(GL)		Maximum L	(HuL_1)
Upper Facial Ht.	(G'H)		RADIUS	
Bimaxillary B	(GB)		Maximum L	(RaL_1)
Nasal Ht.	(NH')		ULNA	
Nasal B	(NB)		Maximum L	(UlL_1)

N.B. The following symbols are an aid to quick notation:

/ = present (coded 1)

X = absent (coded 0)

- - = not possible to ascertain owing to condition of bone

a.m.; p.m. = ante-mortem; post-mortem.

assign it to a particular disease. In the case of joint disease, it is worth noting the changes in each bone instead of giving a general comment on its presence.

Measurements should be taken carefully and with full understanding of the points involved. In all cases, the recordings should be at least to the nearest millimetre and in the case of small dimensions to 0.1 mm. Although only basic measurements are required in the blank, it is often valuable to take others if time allows. If there is any suspicion that an instrument is inaccurate, it should be abandoned at once.

Now, is any highly intricate analysis expected in short primary osteological reports? The answer is no! Only in the case of large samples (preferably over thirty specimens for each sex) is it of value to compute means, standard deviations and so forth. In any case this is work for a specialist and is beyond the scope of the shorter report.

Comments on certain general morphological features ('anthroposcopic' data) are perhaps of least value in works on post-Mesolithic skeletons. This category includes the shape of the vault (seen from the upper or *norma verticalis* view), the shape of the orbits or chin, and so forth. The degree of prominence of such features as the supra-orbital ridges or mastoid processes are of course important in regard to sexing, but when noting the exact degree of development difficulties arise. It is extremely difficult to know whether each worker is classifying the specimens in exactly the same way; for example, one person may consider the vault shape as 'ellipsoid' while another might regard it as 'ovoid'. It therefore seems wiser to leave out this form of description unless a particularly abnormal shape seems worthy of comment.

Since the first edition of this handbook, there has been a growing interest on the part of archaeologists in the information to be derived from the study of skeletal material. In Britain, at least, various excavators have taken on themselves the preliminary preparation and analysis of the burials they have found. It is to be hoped that this revised handbook will continue to encourage the excavators of ancient human remains to take a further positive interest in them.

VIII   Bibliography

i) Some standard works of reference

Anatomy of the skeleton

BREATHNACH, A. S. (ed.) 1958. *Frazer's Anatomy of the Human Skeleton.* 247 pp. London.

ECKHOFF, N. L. 1946. *Aids to Osteology,* 4th. edn. 260 pp. London.

GRANT, J. C. B. 1968. *An Atlas of Anatomy,* 5th edn. London.

JAMIESON, E. B. 1937. *Dixon's Manual of Human Osteology.* 465 pp. London.

JOHNSTON, T. B. *et al.* (eds). 1958. *Gray's Anatomy.* 1604 pp. London.

WOERDEMAN, M. W. 1948. *Atlas of Human Anatomy, I. Osteology, Arthrology, Myology.* Amsterdam.

Anatomy of bone

LE GROS CLARK, W. E. 1958. *The Tissues of the Body.* 203 pp. Oxford.

MURRAY, P. D. F. 1936. *Bones.* 203 pp. Cambridge.

WEINMANN, J. P. & SICHER, H. 1947. *Bone and Bones. Fundamentals of Bone Biology.* 464 pp. London.

Sex, age and stature

CAMPS, F. E. & PURCHASE, W. B. 1956. *Practical Forensic Medicine.* 541 pp. London.

STEWART, T. D. & TROTTER, M. 1954. *Basic Readings on the Identification of Human Skeletons: Estimation of Age.* 347 pp. New York.

Osteometry

ASHLEY MONTAGU, M. F. 1951. *An Introduction to Physical Anthropology.* 555 pp. Illinois.

COMAS, J. 1957. *Manual de Antropología Física.* 698 pp. Buenos Aires (Fondo de Cultura Económica).

HOOTON, E. A. 1947. *Up from the Ape.* 788 pp. New York.

HOWELLS, W. W. 1975. Cranial variation in Man. A study by multivariate analysis of patterns of difference among Recent human populations. *Pap. Peabody Mus.,* **67**. 259 pp.

MARTIN, R. & SALLER, K. 1956/1959. *Lehrbuch der Anthropologie,* I and II. 1574 pp. Stuttgart.

MUKHERJEE, R., RAO, C. R. & TREVOR, J. C. 1955. *The Ancient Inhabitants of Jebel Moya (Sudan).* 123 pp. Cambridge.

STEWART, T. D. (ed.) 1947. *Hrdlička's Practical Anthropometry,* 3rd. ed. 230 pp. Philadelphia, Wistar Institute.

Jaws and teeth

BRASH, J. C. *et al.* 1956. *The Aetiology of Irregularity and Malocclusion of the Teeth.* 241 pp. London (Dental Board of U.K.).

BROTHWELL, D. R. (ed.) 1963. *Dental Anthropology.* 288 pp. London.

COLYER, J. F. & SPRAWSON, E. 1946. *Dental Surgery and Pathology.* 1067 pp. London.

KLATSKY, H. & FISHER, R. L. 1953. *The Human Masticatory Apparatus.* 246 pp. London.

MILES, A. E. W. (ed.) 1967. *Structural and Chemical Organization of Teeth.* 2 vols, 525 + 489 pp. London, Academic Press.

SCOTT, J. H. & SYMONS, N. B. B. 1958. *Introduction to Dental Anatomy.* 344 pp. Edinburgh.

Statistics

CHAMBERS, E. G. 1955. *Statistical Calculation for Beginners.* 168 pp. Cambridge.

DAHLBERG, G. 1948. *Statistical Methods for Medical and Biological Students.* 139 pp. London.

FISHER, R. A. 1954. *Statistical Methods for Research Workers.* 356 pp. Edinburgh.

MORONEY, M. J. 1954. *Facts from Figures.* 472 pp. London, Penguin Books.

ORTON, C. 1980. *Mathematics in Archaeology.* 248 pp. London.

SEAL, H. L. 1964. *Multivariate Statistical Analysis for Biologists.* 207 pp. London.

Disease

ARMELAGOS, G. J., MIELKE, J. H. & WINTER, J. 1971. Bibliography of Human Paleopathology. *Res. Reps Univ. Mass.*, Amherst, **8**.

BROTHWELL, D. R. & SANDISON, A. T. (eds). 1967. *Diseases in Antiquity*. 766 pp. Springfield.

CRAIN, J. B. 1971. Human Palaeopathology: A Bibliographic List. *Pap. Sacramento anthrop. Soc.*, **12**.

FAIRBANK, T. 1951. *An Atlas of General Affections of the Skeleton*. 411 pp. Edinburgh.

GREIG, D. M. 1931. *Clinical Observations on the Surgical Pathology of Bone*. 248 pp. Edinburgh.

HARE, R. 1954. *Pomp and Pestilence*. 244 pp. London.

HENSCHEN, F. 1966. *The History of Diseases*. 344 pp. London.

HERTZLER, A. E. 1931. *Surgical Pathology of the Diseases of Bones*. 272 pp. Philadelphia.

MOODIE, R. L. 1923. *Palaeopathology: an introduction to the study of ancient evidences of disease*. 567 pp. Illinois.

PALES, L. 1930. *Paléopathologie et Pathologie Comparative*. 352 pp. Paris.

SIGERIST. H. E. 1951. *A History of Medicine, I. Primitive and Archaic Medicine*. 564 pp. New York.

STEINBOCK, R. T. 1976. *Paleopathological Diagnosis and Interpretation. Bone Diseases in Ancient Human Populations*. 440 pp. Springfield.

Radiography

ARCHER, V. W. 1947. *The Osseous System. A Handbook of Roentgen Diagnosis*. 320 pp. Chicago.

BRAILSFORD, J. F. 1953. *The Radiology of Bones and Joints*. 875 pp. London.

MIDDLEMISS, H. 1961. *Tropical Radiology*. 272 pp. London.

SIMON, G. 1965. *Principles of Bone X-ray Diagnosis*. 201 pp. London.

General

BROTHWELL, D. R. (ed.). 1968. *The Skeletal Biology of Earlier Human Populations*. 288 pp. London.

BROTHWELL, D. R. & HIGGS, E. (eds) 1969. *Science in Archaeology*. 720 pp. London.

BROTHWELL, D. R. & CHIARELLI, B. A. (eds). 1973. *Population Biology of the Ancient Egyptians*. London.

CORNWALL, I. W. 1956. *Bones for the Archaeologist*. 255 pp. London.

DINGWALL, E. J. 1931. *Artificial Cranial Deformation*. 313 pp. London.

FINNEGAN, M. & FAUST, M. A. 1974. Bibliography of human and non-human non-metric variation. *Res. Reps Univ. Mass.*, Amherst, **14**. 133 pp + suppl. 13 pp.

KROGMAN, W. M. 1962. *The Human Skeleton in Forensic Medicine*. 337 pp. Springfield.

STANILAND, L. N. 1952. *The Principles of Line Illustration*. 212 pp. London.

ii) References to literature quoted in the text

ABELSON, P. H. 1954. Palaeobiochemistry. *Yb. Carnegie Inst. Washington, 53*: 97–101.

ACHESON, R. M., HEWITT, D., WESTROPP, C., & MCINTYRE, M. N. 1956. Some effects of adverse environmental circumstances on skeletal development. *Am. J. phys. Anthrop.*, Washington, (n.s.) **14**: 375.

ACKERKNECHT, E. H. 1953. Palaeopathology; a survey. In *Anthropology Today*: 120 127 Chicago.

ACSÁDI, G. & NEMESKERI, J. 1970. *History of Human Life Span and Mortality*. 346 pp. Budapest, Akadémai Kiadó.

AICHEL, O. 1915. Die normale Entwicklung der Schuppe des Hinterhauptsbeines, die Entstehung der 'Inkabein' genannten Anomalie der Schuppe und die Kausale Grundlage für die typischen Knochennähte an der Schuppe. *Archiv. Anthrop.*, Braunschweig, **13**: 130–168.

AKABORI, E. 1933. *Jap. J. Med. Sci.*, **4**: 61–315. (Quoted by Laughlin & Jørgensen 1956).

ALLBROOK, D. B. 1955. The East African vertebral column. A Study in racial variability. *Am. J. phys. Anthrop.*, Washington, (n.s.) **13**: 489–511.

ANDERMANN, S. 1976. The cnemic index: a critique. *Am. J. phys. Anthrop.*, Washington, **44**: 369-370.

ANDERSEN, J. G. 1969. *Studies in the Mediaeval diagnosis of leprosy in Denmark.* 142 pp. Costers, Copenhagen.

ANDERSON, J. E. 1963. The people of Fairty. An osteological analysis of an Iroquois ossuary. *Bull. natn. Mus. Can.*, Ottawa, **193**: 28–129.

—— 1968. Skeletal 'anomalies' as genetic indicators. *In* Brothwell, D. R. (ed.), *The skeletal biology of earlier human populations:* 135–147. London, Pergamon.

ANDREWS, P. & WILLIAMS, D. B. 1973. The use of principal components analysis in physical anthropology. *Am. J. phys. Anthrop.*, Washington, **39**: 291–304.

ANGEL, J. L. 1943. Treatment of archaeological skulls. *In* Shapiro, H. L. (ed.), *Anthrop. Briefs*, New York, **3**: 3–8.

—— 1946. Skeletal change in ancient Greece. *Am. J. phys. Anthrop.*, Washington, (n.s.) **4**: 69–97.

—— 1950. Skeletons. *Archaeology*, **3**: 233–241.

—— 1964. Osteoporosis: thalassemia? *Am. J. phys. Anthrop.*, Washington, **22**: 369–374.

—— 1966. Porotic hyperostosis, anemias, malarias, and marshes in the prehistoric eastern Mediterranean. *Science*, **153**: 760–763.

—— 1969. The bases of paleodemography. *Am. J. phys. Anthrop.*, Washington, **30**: 427–438.

—— & COON, C. S. 1954. La Cotte de St. Brelade II: present status. *Man*, London, **54**: 53–55.

ARMOUR-CLARK, H. 1957. Radiographical duplication of radiographs. *Brit. dent. J.*, London, **102**: 299–304.

ASCENZI, A. 1955. Some histochemical properties of the organic substance in Neanderthal bone. *Am. J. phys. Anthrop.*, Washington, (n.s.) **13**: 557–566.

—— 1969. Microscopy and prehistoric bone. *In* Brothwell, D. & Higgs, E. (eds), *Science in Archaeology*: 526–538, London, Thames & Hudson.

ASHLEY, G. T. 1956. The human sternum: the influence of sex and age on its measurements. *J. for Med.*, Cape Town, **3**: 27–43.

ASHLEY MONTAGU, M. F. 1933–34. The anthropological significance of the pterion in the primates. *Am. J. phys. Anthrop.*, Washington, (n.s.) **18**: 159–336.

—— 1937. The medio-frontal suture and the problem of metopism in the primates. *J. Roy. anthrop. Inst.*, London, **67**: 157–201.

—— 1951. *An Introduction to Physical Anthropology.* 555 pp. Illinois.

BAINBRIDGE, D. & GENOVÉS, S. T. 1956. A study of sex differences in the scapula. *J. Roy. anthrop. Inst.*, London, **86**: 109–134.

BANKS, H. V. 1934. Incidence of third molar development. *Angle Orthodont.*, Chicago, **4**: 223–233.

BARKER, H. & MACKEY, J. 1961. British Museum National Radiocarbon Measurements, III. *Radiocarbon*, New Haven, **3**: 39–45.

BARNICOT, N. A. & BROTHWELL, D. R. 1959. The evaluation of metrical data in the comparison of ancient and modern bones. In *Medical Biology and Etruscan Origins:* 131–149. London.

BATRAWI, A. & MORANT, G. M. 1947. A study of the First Dynasty series of Egyptian skulls from Sakkara and of an Eleventh Dynasty series from Thebes. *Biometrika*, Cambridge, **34**: 18–27.

BAUME, R. M. & CRAWFORD, M. H. 1978. Discrete dental traits in four Tlaxcaltecan Mexican populations. *Am. J. phys. Anthrop.*, Washington, **49**: 351–360.

BENINGTON, R. CREWDSON 1912. A study of the Negro skull with special reference to the Congo and Gaboon crania. *Biometrika*, Cambridge, **8**: 292–339.

BENNINGHOFF, A. 1925. Spaltlinien am Knochen, eine Methode zur Ermittlung der Architectur platter Knocken. *Verh. anat. Ges.*, Jena, **34**: 189–206.

BERGMANN, R. A. M. & KARSTEN, P. 1952. The fluorine content of *Pithecanthropus* and of other specimens from the Trinil Fauna. *Proc. Koninkl. Nederl. Akad. Wetenschappen*, **55**: 150–152.

BENNETT, K. A. 1965. The etiology and genetics of wormian bones. *Am. J. phys. Anthrop.*, Washington, (n.s.) **23**: 255–260.

BERRY, A. C. 1974. The use of non-metrical variations of the cranium in the study of Scan-
dinavian population movements. *Am. J. phys.
Anthrop.*, Washington, **40**: 345–358.

—— 1975. Factors affecting the incidence of
non-metrical skeletal variants. *J. Anat.*,
London, **120**: 519–535.

—— 1978. Anthropological and family studies
on minor variants of the dental crown. *In*
Butler, P. M. & Joysey, K. A. (eds), *Development, function and evolution of teeth:* 81–98.
London.

—— & BERRY, R. J. 1967. Epigenetic variation in
the human cranium. *J. Anat.*, London, **101**:
361–379.

——, —— & UCKO, P. J. 1967. Genetical change in
ancient Egypt. *Man*, **2**: 551–568.

BERRY, R. J. 1968. The biology of non-metrical
variation in mice and men. In *The Skeletal
Biology of Earlier Human Populations:*
103–133. London.

BIRKBY, W. H. 1966. An evaluation of race and sex
identification from cranial measurements.
Am. J. phys. Anthrop., Washington, (n.s.) **24**:
21–28.

BLACK, T. K. 1978. A new method for assessing the
sex of fragmentary skeletal remains: femoral
shaft circumference. *Am. J. phys. Anthrop.*,
Washington, **48**: 227–232.

BLACKWOOD, B. & DANBY, P. M. 1955. A study of
artificial cranial deformation in New Britain.
J. Roy. Anthrop. Inst., London, **85**: 173–191.

BLUMBERG, J. M. & KERLEY, E. 1966. A critical consideration of roentgenology and microscopy
in palaeopathology. *In* Jarcho, S. (ed.),
Human Palaeopathology: 150–170. New
Haven, Yale University Press.

BOESSNEK, J. 1969. Osteological differences
between sheep and goat. In *Science in
Archaeology:* 331–358. London.

BORGOGNINI, S. M. 1968*a*. Studio di alcune carat-
teristiche del siero emoagglutinante anti-H,
estratto dai semi di *Ulex europaeus*, usato
nella determinazione dei gruppi sanguigni
ABO in ossa umane recenti ed antiche. II.
Memorie Soc. tosc. Sci. Nat. B, **75**: 21–30.

—— 1968*b*. Dimostrazione della presenza di
alcuni costituenti fondamentali delle sostanze
gruppo-specifiche ABO in ossa di diversa
antichita. *Memorie Soc. tosc. Sci. Nat. B*, **75**:
202–217.

—— & PAOLI, G. 1969. Biochemical and
immunological investigations on early
Egyptian remains. *J. hum. Evol.* **1**: 281–287.

BOSTANCI, E. Y. 1959. The astragalus and
calcaneus of the Roman people of Gordion in
Anatolia. *Belleten*, Ankara, **23**: 177–201.

BOUVEIR, M. & UBELAKER, D. H. 1977. A comparison
of two methods for the microscopic determination of age at death. *Am. J. phys.
Anthrop.*, Washington, **46**: 391–394.

BOYD, J. D. & TREVOR, J. C. 1953. Problems in recon-
struction. *In* Simpson, K. (ed.), *Modern
Trends in Forensic Medicine:* 133–152.
London.

BOYD, W. C. 1950. *Genetics and the Races of Man.*
453 pp. Oxford.

—— & BOYD. L. G. 1933. Blood grouping by
means of preserved muscle. *Science*, New
York, **78**: 578.

BRASH, J. C. 1956. *The Aetiology of Irregularity
and Malocclusion of the Teeth*, **1**. 241 pp.
London (Dental Board of the U.K.).

BREITINGER, E. 1937. Zur Berechnung der
Körperhöhe aus den langen Gliedmassen
Knocken. *Anthrop. Anz.*, Stuttgart, **14**:
249–274.

BRIGGS, L. C. 1955. The Stone Age Races of
Northwest Africa. *Bull. Am. Sch. Prehist.
Res.*, New Haven, **18**: 1–98.

—— 1958. *Initiation à l'Anthropologie du
Squelette.* 56 pp. Algiers.

BROCKMAN, E. P. 1948. Diseases of bone. *In*
Handfield-Jones, R. M. & Porritt, A. E. (eds),
The Essentials of Modern Surgery:
1056–1101. Edinburgh.

BRODRICK, A. H. 1948. *Early Man; a Survey of
Human Origins.* 288 pp. London.

BROEK, A. J. P. VAN DEN 1943. On exostoses in the
human skull. *Acta Neerl. Morph.*, Utrecht, **5**:
95–118.

BROMAN, G. E. 1957. Precondylar tubercles in
American Whites and Negroes. *Am. J. phys.
Anthrop.*, Washington, (n.s.) **15**: 125–135.

BRONGERS, J. A. 1966. Ultra-violet fluorescence
photography of a soil silhouette of an interred
corpse. *Ber. Rijksdienst oudheidk. Bodemon-
derz.* **15–16**: 227–228.

BROOKS, S. T. 1955. Skeletal age at death; the
reliability of cranial and pubic age indicators
Am. J. phys. Anthrop., Washington, (n.s.) **13**:
567–590.

—— & MELBYE, J. 1967. Skeletal lesions suggestive of Pre-Columbian multiple myeloma in a burial from the Kane Mounds, near St Louis, Missouri. *Misc. Pap. Paleopath., Mus. N. Arizona*, **1**: 23–29.

BROTHWELL, D. R. 1958*a*. Congenital absence of the basi-occipital in a Romano-Briton. *Man*, London, **58**: 73–74.

—— 1958*b*. Evidence of leprosy in British archaeological material. *Med. Hist.*, London, **2**: 287–291.

—— 1959*a*. The use of non-metrical characters of the skull in differentiating populations. In *Ber. 6 Tag. dtsch. Ges. Anthrop. Kiel:* 103–109. Göttingen.

—— 1959*b*. A rare dental anomaly in archaeological material. *Brit. dent. J.*, London, **107**: 400–401.

—— 1959*c*. Teeth in earlier human populations. *Proc. Nutrit. Soc.*, **18**: 59–65.

—— 1960*a*. The Bronze Age people of Yorkshire: a general survey. *Advmt Sci.*, Lond. **16**: 311–322.

—— 1960*b*. A possible case of mongolism in a Saxon population. *Ann. Hum. Genet.*, London, **24**: 141–150.

—— 1961*a*. An Upper Palaeolithic skull from Whaley Rock Shelter No. 2, Derbyshire. *Man*, London, **61**: 113–116.

—— 1961*b*. The palaeopathology of early British man: an essay on the problems of diagnosis, and analysis. *J. Roy. Anthrop. Inst.*, London, **91**: 318–344.

—— 1961*c*. Cannibalism in early Britain. *Antiquity*, Cambridge, **35**: 304–307.

—— 1963. The macroscopic dental pathology of some earlier human populations. *In* Brothwell, D. R. (ed.), *Dental Anthropology:* 271–288. London, Pergamon.

—— 1965. The palaeopathology of Early-Middle Bronze Age remains from Jericho. *In* Kenyon, K. M., *Jericho*, **2**: 685–693. London.

—— 1967*a*. The evidence for neoplasms. *In* Brothwell, D. & Sandison, A. T., *Diseases in Antiquity*: 320–345. Springfield.

—— 1967*b*. Biparietal thinning in early Britain. *In* Brothwell, D. & Sandison, A. T., *Diseases in Antiquity*: 413–416. Springfield.

—— 1970. The real history of syphilis. *Science Journal*, London, **6**: 27–33.

—— 1971. Palaeodemography. *In* Brass, W. (ed.), *Biological Aspects of Demography*: 111–130. London, Taylor & Francis.

—— 1972. Palaeodemography and earlier British populations. *Wld Archaeol.* **4**: 75–87.

—— 1975. Possible evidence of a cultural practice affecting head growth in some late Pleistocene east Asian and Australasian populations. *J. archaeol. Sci.*, **2**: 75–77.

—— 1978. Possible evidence of the parasitisation of early Mexican communities by the micro-organism *Treponema. Bull Inst. Archaeol.*, **15**: 113–130.

—— & BURLEIGH, R. 1975. Radiocarbon dates and the history of treponematoses in Man. *J. archaeol. Sci.*, **2**: 393–396.

——, CARBONELL, V. M. & GOOSE, D. H. 1963. Congenital absence of teeth in human populations. *In* Brothwell, D. R. (ed.), *Dental Anthropology:* 179–190. London.

—— & KRZANOWSKI, W. 1974. Evidence of biological differences between early British populations from neolithic to medieval times, as revealed by eleven commonly available cranial vault measurements. *J. archaeol. Sci.*, **1**: 249–260.

—— & MØLLER-CHRISTENSEN, V. 1963. Medico-historical aspects of a very early case of mutilation. *Dan. med. Bull.*, **10** (1): 21–25.

——, MOLLESON, T., GRAY, P. & HARCOURT, R. 1969. The application of X-rays to the study of archaeological materials. *In* Brothwell, D. R. & Higgs, E. (eds), *Science in Archaeology*: 513–525. London, Thames & Hudson.

——, —— & METREWELI, C. 1968. Radiological aspects of normal variation in earlier skeletons: an exploratory study. *In* Brothwell, D. R. (ed.), *The skeletal biology of earlier human populations:* 149–172. London.

—— & SANDISON, A. T. (eds). 1967. *Diseases in Antiquity*. 766 pp. Springfield.

—— & SPEARMAN, R. 1963. The hair of earlier peoples. *In* Brothwell, D. R. & Higgs, E. (eds), *Science in Archaeology*: 426–436. London.

BOURKE, J. B. 1967. A review of the palaeopathology of the arthritic diseases. *In* Brothwell, D. R. & Sandison, A. T. (eds), *Diseases in Antiquity:* 352–370. Springfield.

BUCKLAND-WRIGHT, J. C. 1970. A radiographic examination of frontal sinuses in early British populations. *Man*, **5**: 512–517.

BUIKSTRA, J. E. 1977. Differential diagnosis: an epidemiological model. *Yb. phys. Anthrop.*, New York, (1976) **20**: 316–328.

—— & COOK, D. C. 1978. Pre-Columbian tuberculosis: an epidemiological approach. *Med. Coll. Virginia Q.* **14**: 32–44.

BULLEN, A. K. 1972. Palaeoepidemiology and distribution of prehistoric treponemiasis (syphilis) in Florida. *Fla Anthrop.*, **35**: 133–174.

BUXTON, L. H. D. 1938. Playtmeria and platy-cnemia. *J. Anat.*, London, **73**: 31–36.

—— & MORANT, G. M. 1933. The essential craniological technique. Part 1. Definitions of points and planes. *J. Roy. Anthrop. Inst.*, London, **63**: 16–47.

CAMERON, J. 1934. *The Skeleton of British Neolithic Man.* 272 pp. London.

CAMPBELL, T. D. 1939. Food, food values and food habits of the Australian Aborigines in relation to their dental conditions. *Aust. J. Dent.*, Melbourne, **43**: 1, 45, 73, 141, 177.

CAMPS, F. E. 1953. *Medical and Scientific Investigations in the Christie Case.* 244 pp., London.

CANDELA, P. B. 1936. Blood group reactions in ancient human skeletons. *Am. J. phys. Anthrop.*, Washington, **21**: 429–432.

CARBONELL, V. M. 1963. Variations in the frequency of shovel-shaped incisors in different populations. *In* Brothwell, D. R. (ed.), *Dental Anthropology:* 211–234. London.

CARLSON, D. S., ARMELAGOS, G. J. & VAN GERVEN, D. P. 1974. Factors influencing the etiology of cribra orbitalia in prehistoric Nubia. *J. hum. Evol.*, **3**: 405–410.

—— —— —— 1976. Patterns of age-related cortical bone loss (osteoporosis) within the femoral diaphysis. *Hum. Biol.*, **48**: 295–314.

CARROLL, D. S. 1957. Roentgen manifestations of sickle cell disease. *Sth. med. J.*, La Grange, **50**: 1486–1490.

CAVE, A. J. E. 1956. Cremated remains from Mound 11. Appendix III *in* The Pagan-Danish Barrow Cemetery at Heath Wood, Ingleby. 1955 Excavations. *J. Derbysh. Archaeol. nat. Hist. Soc.*, **76**: 55.

CHAPMAN, F. H. 1972. Vertebral osteophytosis in prehistoric populations of central and southern Mexico. *Am. J. phys. Anthrop.*, Washington, **36**: 31–38.

CLEMENT, A. J. 1956. Caries in the South African ape-man: some examples of undoubted pathological authenticity believed to be 800,000 years old. *Brit. dent. J.*, London, **1956**: 4–7.

CLEMOW, F. G. 1903. *The Geography of Disease.* 624 pp. Cambridge.

COBB, M. W. 1952. Skeleton. *In* Lansing, A. I. (ed.), *Cowdry's Problems of Ageing*, 2nd. edn. 1061 pp. Baltimore.

—— 1955. The age incidence of suture closure (Abstract). *Am. J. phys. Anthrop.*, Washington, (n.s.) **13**: 394.

COLYER, F. 1936. *Variations and Diseases of the Teeth of Animals.* 750 pp. London.

COMAS, J. 1957. *Manual de Antropologica Física.* 698 pp. Buenos Aires.

CONSTANDSE-WESTERMANN, T. S. 1974. L'homme Mésolithique du nord-ouest de l'Europe, distances biologiques, considérations généti-ques. *Bull. Mém. Soc. Anthrop. Paris*, **13**: 173–199.

COOKSON, M. B. 1954. *Photography for Archaeologists.* 123 pp. London.

CORNWALL, I. W. 1954. The human remains from Sutton Walls. *Arch. J.*, London, **60**: 66–78.

—— 1956. *Bones for the Archaeologist.* 255 pp. London.

COURVILLE, C. B. 1950. Cranial injuries in pre-historic man with particular references to the Neanderthals. *Yb. phys. Anthrop.*, **1950**: 185–205.

COUTURIER, P. 1959. L'énigme des tori maxillaires: torus palatinus et torus mandibularis. Contribution à leur étude radio-clinique et pathogénique. *Revue fr. Odonto-Stomat.*, **3**: 1–22.

CRICHTON, J. M. 1966. A multiple discriminant analysis of Egyptian and African Negro crania. *Papers Peabody Mus. Arch. Ethnol.*, **57**: 47–67.

CRUMP, J. A. 1901. Trephining in the South Seas. *J. Roy. Anthrop. Inst.*, London, **31**: 167–172.

CUNNINGHAM, D. J. 1951. *Textbook of Anatomy* (ed. Brash, J. C.). 1604 pp. Oxford.

CUNNINGTON, M. E. 1949. *An Introduction to the archaeology of Wiltshire from the earliest times to the Pagan Saxons.* 172 pp. Devizes.

CYBULSKI, J. S. 1977. Cribra orbitalia, a possible sign of anemia in historic native populations of the British Columbia coast. *Am. J. phys. Anthrop.*, Washington, 47: 31–40.

CZARNETZKI, A. 1971–72. Epigenetische Skelettmerkmale im populationsvergleich. *Z. Morph. Anthrop.*, 63: 238–254; 64: 145–158.

DAHLBERG, A. A. 1945. Paramolar tubercle (Bolk). *Am. J. phys. Anthrop.*, Washington, (n.s.) 3: 97–103.

—— 1951. The dentition of the American Indian. *In: The Physical Anthropology of the American Indian*: 138–176. New York.

DANIELSON, K. 1970. Odontodysplasia leprosa in Danish medieval skeletons. *Tandlægebladet*, 74: 603–625.

DAVALOS, E. 1970. Pre-hispanic osteopathology. *In* Stewart, T. D., (ed.), *Handbook of Middle American Indians*, 9 (Physical Anthropology): 68–81. Austin.

DAY, M. H. & MOLLESON, T. I. 1973. The Trinil femora. *In* Day, M. H., (ed.), *Human Evolution*: 127–154. London, Taylor & Francis.

—— & PITCHER-WILMOTT, R. W. 1975. Sexual differentiation in the innominate bone studied by multivariate analysis. *Ann. Hum. Biol.*, 2: 143–151.

DE SOUZA, P. & HOUGHTON, P. 1977. The mean measure of divergence and the use of non-metric data in the estimation of biological distances. *J. archaeol. Sci.*, 4 (2): 163–169.

DE TERRA, H. 1957. *Man and Mammoth in Mexico.* 191 pp. London.

DE VRIES, H. & OAKLEY, K. P. 1959. Radiocarbon dating of the Piltdown skull and jaw. *Nature, Lond.*, 184: 224–226.

DENNINGER, H. S. 1931. Cervical ribs: a prehistoric example. *Am. J. phys. Anthrop.*, Washington, 16: 211–215.

DENNISON, J. 1979. Citrate estimation as a means of determining the sex of human skeletal material. *Archaeol. phys. Anthrop. Oceania*, 14: 136–143.

DERRY, D. E. 1913. A case of hydrocephalus in an Egyptian of the Roman period. *J. Anat. Lond.*, 48: 436–458.

DEWEY, J. R., ARMELAGOS, G. J. & BARTLEY, M. H. 1969. Femoral cortical involution in three Nubian archaeological populations. *Hum. Biol.*, 41: 13–28.

DICK, J. L. 1922. *Rickets.* 488 pp. London.

DICKSON, G. C. 1970. The natural history of malocclusion. *Dent. Practnr dent. Rec.*, 20: 216–232.

DINGWALL, E. J. 1936. *Artificial Cranial Deformation; a Contribution to the Study of Ethnic Mutilations.* 313 pp. London.

DITCH, L. E. & ROSE, J. C. 1972. A multivariate dental sexing technique. *Am. J. phys. Anthrop.*, Washington, 37: 61–64.

DIXON, A. F. 1900. On certain markings on the frontal part of the human cranium and their significance. *J. Roy. Anthrop. Inst.*, London, 30: 96–97.

DOBROVSKY, M. 1946. Abrasiones dentarias en cráneos de Indios Patagones. *Rev. Mus. La Plata* (n.s., Secc. Antropologia) 2: 301–347.

DODO, Y. 1974. Non-metrical cranial traits in the Hokkaido Ainu and the Northern Japanese of recent times. *J. anthrop. Soc. Nippon*, 82: 31–51.

DORSEY, G. A. 1897. A Maori skull with double left parietal bone. *Chicago Med. Rec.*, 12: 1–4.

DREXLER, L. 1957. Ein pathologischer Humerus eines Höhlenbären. *Ann. naturh. Mus. Wien.*, 61: 96–101.

DUCKWORTH, W. L. H. 1915. *Morphology and Anthropology.* 564 pp. Cambridge.

DUPERTUIS, C. W. & HADDEN, J. A. 1951. On the reconstruction of stature from long-bones. *Am. J. phys. Anthrop.*, Washington, (n.s.) 9: 15–53.

DUTRA, F. R. 1944. Identification of person and determination of cause of death from skeletal remains. *Arch. Pathol.*, 38: 339.

DWIGHT, T. 1890. The closure of the cranial sutures as a sign of age. *Boston Med. Surg. J.*, 122: 1–12.

ECKHOFF, N. L. 1946. *Aids to Osteology.* 4th ed. 260 pp. London.

ELLIOT SMITH, G. 1908. The most ancient splints. *Brit. med. J.*, London, 1: 732–773.

—— & DAWSON, W. R. 1924. *Egyptian Mummies.* 190 pp. London.

—— & WOOD JONES, F. 1910. Report on the human remains. *Archaeological survey of Nubia. Report of 1907-1908*, 11. 375 pp. (Issued by Ministry of Finance, Egyptian Survey Dept., Cairo.)

EL-NAJJAR, M. Y. & DAWSON, G. L. 1977. The effect of artificial cranial deformation on the incidence of wormian bones in the lambdoidal suture. *Am. J. phys. Anthrop.*, Washington, **46**: 155–160.

ESCOBAR, V., MELNICK, M. & CONNEALLY, P. M. 1976. The inheritance of bilateral rotation of maxillary central incisors. *Am. J. phys. Anthrop.*, Washington, **45**: 109–116.

EVANS, F. G. & GOFF, C. W. 1957. A comparative study of the primate femur by means of the stresscoat and the split-line techniques. *Am. J. phys. Anthrop.*, Washington, **15**: 59–89.

FAIRBANK, T. 1951. *An Atlas of the General Affections of the Skeleton.* 411 pp. Edinburgh.

FALKNER, F. 1957. Deciduous tooth eruption. *Arch. Diseases of Childhood*, London, **32**: 386–391.

FELL, C. 1956. Roman burials found at Arbury Road, Cambridge, 1952. *Proc. Camb. Antiq. Soc.*, **49**: 13–23.

FINNEGAN, M. 1978. Non-metric variation of the infracranial skeleton. *J. Anat.*, **125**: 23–37.

—— & COOPRIDER, K. 1978. Empirical comparison of distance equations using discrete traits. *Am. J. phys. Anthrop.*, Washington, **49**: 39–46.

—— & FAUST, M. A. 1974. Bibliography of human and non-human non-metric variation. *Res. Reps Univ. Mass.*, Amherst, **14**. 133 pp + suppl. 13 pp.

FISHER, R. A. 1936. Coefficient of racial likeness and the future of craniometry. *J. Roy. Anthrop. Inst.*, London, **46**: 57–63.

FLANDER, L. B. 1978. Univariate and multivariate methods of sexing the sacrum. *Am. J. phys. Anthrop.*, Washington, **49**: 103–110.

FLOWER, W. H. 1898. *Essays on Museums.* 394 pp. London.

FORBES, G. 1941. The effects of heat on the histological structure of bone. *Police J.*, **14** (1): 50–60.

FRAYER, D. W. 1978. The evolution of the dentition in Upper Palaeolithic and Mesolithic Europe. *Univ. Kansas Publs Anthrop.*, Lawrence, **10**. 201 pp.

FURNEAUX, W. S. 1895. *Human Physiology.* 249 pp. London.

FUSTÉ, M. 1955. Antropología de las poblaciones pirenáicas durante el período neo-eneolítico. *Trab. Inst. 'Bernardino de Sahagún' Antrop. y Etnología*, **14**: 109–135.

GARCÍA FRÍAS, E. 1940. La tuberculosis en los antiguos Peruanos. *Actualidad Médica Peruana*, **5**: 274–291.

GARN, S. M. 1973. Adult bone loss, fracture epidemiology and nutritional implications. *Nutrition*, **27**: 107–115.

—— & HELMRICH, R. H. 1967. Next step in automated anthropometry. *Am. J. phys. Anthrop.*, Washington, **26**: 97–100.

——, LEWIS, A. B. & KEREWSKY, R. S. 1967. Sex difference in tooth shape. *J. dent. Res.*, **46**: 1470.

——, —— & POLACHEK, D. L. 1959. Variability of tooth formation. *J. dent. Res.*, Baltimore, **38**: 135–148.

——, ——, SWINDLER, D. R. & KEREWSKY, R. S. 1967. Genetic control of sexual dimorphism in tooth size. *J. dent. Res.*, **46**: 963–972.

—— & ROHMANN, C. G. 1966. Interaction of nutrition and genetics in the timing of growth and development. *Pediat. Clins N. Am.*, **13**: 353–379.

——, SILVERMAN, F. N., HERTZOG, K. P. & ROHMANN, C. G. 1968. Lines and bands of increased density. Their implication to growth and development. *Med. Radiogr. Photogr.*, **44**: 58–89.

——, WAGNER, B., ROHMANN, C. G. & ASCOLI, W. 1968. Further evidence for continuing bone expansion. *Am. J. Phys. Anthrop.*, Washington, **28**: 219–222.

GÁSPÁRDY, G. & NEMESKÉRI, J. 1960. Palaeopathological studies on Copper Age skeletons found at Alsónémedi. *Acta Morphol.*, Budapest, **9**: 203–219.

GEJVALL, N.-G. 1947. Bestämning av brända ben fran fortnida gravar. *Fornvännen*, **1**: 39–47.

—— 1951. Gravfältet i Mellby by Källands härad. II. Antropologisk del *Västergötlands Fornminnesforenings Todskrift*, **5** (6): 53–57.

—— 1969. Cremations. *In* Brothwell, D. R. & Higgs, E. (eds), *Science in Archaeology*: 468–479. London.

—— & SAHLSTRÖM, K. E. 1948. Gravfältet på kyrkbacken i Horns socken Västergötland. II. Antropologisk del *Kungl. Vitterhets Historie och Antikvitets Akademiens Handl.*, Del 60, **2**: 153–180.

GENOVÉS, S. 1954. The problem of the sex of certain fossil hominids, with special reference to the Neanderthal skeletons from Spy. *J. Roy. Anthrop. Inst.*, London, **84**: 131–144.

—— 1959. *Diferencias Sexuales en el Hueso Coxal.* 440 pp. Instituto de Historia. Mexico.

—— 1967. Proportionality of the long-bones and their relation to stature among Mesoamericans. *Am. J. phys. Anthrop.,* Washington, 26: 67–78.

—— 1969. Estimation of age and mortality. *In* Brothwell, D. R. & Higgs, E. (eds), *Science in Archaeology:* 440–452. London.

—— & MESSMACHER, M. 1959. Valor de los patrones tradicionales para la determinación del edad por medio de las suturas en cráneos mexicanos. *Cuadernas del Instituto de Historia.,* Serie No. 7: 7–53.

GETZ, B. 1955. The hip joint in Lapps and its bearing on the problem of congenital dislocation. *Acta orthop. Scand.,* Copenhagen, Supplement 18. 81 pp.

GILBERT, B. M. & MCKERN, T. W. 1973. A method for ageing the female *os pubis. Am. J. phys. Anthrop.,* Washington, 38: 31–38.

GILBERT, R. I. 1977. Application of trace element research to problems in archaeology. In *Biocultural Adaption in Prehistoric America*: 85–100. University of Georgia Press, Athens.

GILBEY, B. E. & LUBRAN, M. 1953. The ABO and Rh blood group antigens in Predynastic Egyptian mummies. *Man,* London, 30: 23.

GILES, E. 1964. Sex determination by discriminant function analysis of the mandible. *Am. J. phys. Anthrop.,* Washington, (n.s.) 22: 129–136.

—— & ELLIOT, O. 1962. Race identification from cranial measurements. *J. Forensic Sci.,* 7: 147–157.

—— —— 1963. Sex determination by discriminant function analysis of crania. *Am. J. phys. Anthrop.,* Washington, (n.s.) 21: 53–68.

GLANVILLE, E. V. 1967. Perforation of the coronoid-olecranon septum. Humero-ulnar relationships in Netherlands and African populations. *Am. J. phys. Anthrop.,* Washington, 26: 85–92.

GOLDSTEIN, M. S. 1943. *Demographic and bodily changes in descendents of Mexican immigrants.* Austin, Texas (Inst. Latin-American Studies, Univ. Texas).

—— 1948. Dentition of Indian crania from Texas. *Am. J. phys. Anthrop.,* Washington, (n.s.) 6: 63–84.

—— 1957. Skeletal pathology of early Indians in Texas. *Am. J. phys. Anthrop.,* Washington, (n.s.) 15: 299–311.

GOODMAN, C. N. & MORANT, G. M. 1940. The human remains of the Iron Age and other periods from Maiden Castle, Dorset, *Biometrika,* Cambridge, 31: 295–312.

GOODWIN, A. J. H. 1945. *Method in Prehistory.* (South African Archaeological Journal, Handbook Series. No. 1). 191 pp. Cape Town.

GOULD, S. J. 1975. On the scaling of tooth size in mammals. *Am. Zool.,* 15: 351–362.

GRANDJEAN, P. & HOLMA, B. 1973. A history of lead retention in the Danish population. *Environl Physiol. Biochem.,* 3: 268–273.

GRAY, M. 1958. A method for reducing non-specific reactions in the typing of human skeletal material. *Am. J. phys. Anthrop.,* Washington, (n.s.) 16: 135–139.

GREEN, R. F. & SUCHEY, J. M. 1976. The use of inverse sine transformations in the analysis of non-metric cranial data. *Am. J. phys. Anthrop.,* Washington, 45: 61–68.

GREULICH, W. W. & PYLE, S. I. 1950. *Radiographic Atlas of Skeletal Development of the Hand and Wrist.* 190 pp. Stanford.

GRIEG, D. M. 1926. Oxycephaly. *Edinb. Med. J.,* 33: 189–218; 280–302; 357–376.

—— 1931. *Clinical Observations on the Surgical Pathology of Bone.* 248 pp. Edinburgh.

GRIMM, H. 1972. Materialzuwachs und Ideenfortschritt in der Wirbelsäulenforschung an ur- und frühgeschichtlichem sowie mittelalterlichem und frühneuzeitlichem Material. *Homo,* 23: 74–80.

—— & PLATHNER, C. H. 1952. Über ein jungensteinzeitlichen Hydrocephalus von Seeburg in Mansfelder Seekreis und sein Gebiss. *Dtsch. Zahn-, Mund-, und Kieferheilkunde,* 15 (11–12): 1–7.

GUSTAFSON, G. 1950. Age determinations on teeth. *J. Amer. Dent. Assoc.,* Chicago, 41: 45–54.

HACKETT, C. J. 1936. Boomerang leg and yaws in Australian aborigines. *Trans. R. Soc. trop. Med. Hyg.,* 30: 137–150.

—— 1963. On the origin of the human treponematoses. *Bull. Wld Hlth Org.,* 29: 7–41.

—— 1967. The human treponematoses. In *Diseases in Antiquity:* 152–169 Springfield.

—— 1976. *Diagnostic criteria of Syphilis, Yaws and Treponarid (Treponematoses) and of some other diseases in dry bones (for use in osteoarchaeology).* 134 pp. Berlin, Springer-Verlag.

HAMBLY, W. D. 1947. Cranial capacities; a study in methods. *Fieldiana, Anthrop.,* Chicago, **36**: 25–75.

HANIHARA, K. 1967. Racial characteristics in the dentition. *J. dent. Res.,* Suppl. **5**: 923–926.

——, TAMADA, M. & TANAKA, T. 1970. Quantitative analysis of the hypocone in the human upper molars. *J. anthrop. Soc. Nippon,* **78**: 200–207.

HANNA, R. E. & WASHBURN, L. 1953. The determination of the sex of skeletons as illustrated by the Eskimo pelvis. *Hum. Biol.,* Baltimore, **132**: 21–27.

HARE, R. 1954. *Pomp and Pestilence; Infectious Disease, its Origins and Conquest.* 224 pp. London.

HARRIS, H. A. 1933. *Bone growth in Health and Disease.* 248 pp. Oxford.

HARRISON, H. S. 1929. War and the Chase. *Horniman Mus. Handb.,* London, **2631**. 85 pp.

HARRISON, R. J. 1953. *In* Camps, F. E., *Medical and Scientific Investigations in the Christie Case:* 56–99. London, Medical Publications.

HARRISSON, T. 1957. The great cave of Niah: a preliminary report on Bornean prehistory. *Man,* London, **57**: 161–166.

HART, V. L. 1952. *Congenital dysplasia of the hip joint and sequelae.* 187 pp. Illinois.

HEIZER, R. F. & COOK, S. F. 1952. Fluorine and other chemical tests of some North American human and fossil bones. *Am. J. phys. Anthrop.,* Washington, (n.s.) **10**: 289–304.

HELMUTH, H. 1970. Uber den Bau des menschlichen Schadels bei kunstlicher Deformation. *Z. Morph. Anthrop.,* **62**: 30–49.

HENGEN, H. O. 1971. Cribra orbitalia: pathogenesis and probable etiology. *Homo,* **22**: 57–72.

HENKE, W. 1972. Morphometrische untersuchungen am Skelettmaterial des mittelalterlichen Kieler Gertrudenfriedhofs im Vergleich mit anderen nordeuropäischen Skelettserien. *Z. Morph. Anthrop.,* **64**: 308–347.

HEPBURN, D. 1908. Anomalies in the supra-inial portion of the occipital bone, resulting from irregularities of its ossification, with consequent variations of the interparietal bone. *J. Anat. Lond.,* **42**: 88–92.

HERRMAN, B. 1972. Das Combe Capeele-Skelet. *Ausgrabungen Berl.,* **3**: 7–69.

HESS, L. 1946. Ossicula wormiana. *Hum. Biol.,* Baltimore, **18**: 61–80.

HIERNAUX, J. 1963. Heredity and environment: their influence on morphology. A comparison of two independent lines of study. *Am. J. phys. Anthrop.,* Washington, **21**: 575–589.

HILL, K. R., KODIJAT, R. & SARDINI, M. 1951. *Atlas of Framboesia.* 18 pp. Geneva, World Health Organization.

HILLS, M. & BROTHWELL, D. R. 1974. The use of large numbers of variables to measure the shape of a restricted area of bone. *J. archaeol. Sci.,* **1**: 135–150.

HILLSON, S. W. 1979. Diet and dental disease. *Wld Archaeol.,* **11**: 147–162.

HIRSCH, F. & SCHLABOW, K. 1958. Untersuchung von Moorleichenhaaren. *Homo,* **9**: 65–74.

HITCHIN, A. D. & MASON, D. K. 1958. Four cases of compound composite odontomes. *Brit. dent. J.,* London, **104**: 269–274.

HOLT, C. A. 1978. A re-examination of parturition scars on the human female pelvis. *Am. J. phys. Anthrop.,* Washington, **49**: 91–94.

HOOKE, B. G. E. 1926. A third study of the English skull with special reference to the Farringdon Street crania. *Biometrika,* Cambridge, **18**: 1–54.

HOOTON, E. A. 1930. *The Indians of Pecos Pueblo. A Study of Their Skeletal Remains.* 391 pp. New Haven.

—— 1947. *Up from the Ape.* 788 pp. New York.

HOUGHTON, P. 1974. The relationship of the pre-auricular groove of the ilium to pregnancy. *Am. J. phys. Anthrop.,* Washington, **41**: 381–390.

HOWELLS, W. W. 1964. Dctermination du sexe du bassin par fonction discriminate: étude du materiel du docteur Gaillard. *Bull. Mém. Soc. Anthrop. Paris,* **7**: 95–105.

—— 1966. Craniometry and multivariate analysis. *Pap. Peabody Mus.,* **57**: 1–43.

—— 1973. Cranial variation in Man. A study by multivariate analysis of patterns of difference among recent human populations. *Pap. Peabody Mus.,* **67**. 259 pp.

HRDLIČKA, A. 1903. Divisions of the parietal bone in man and other mammals. *Bull. Am. Mus. Nat. Hist.*, New York, **19**: 231–386.

—— 1920. Shovel-shaped teeth. *Am. J. phys. Anthrop.*, Washington, **3**: 429–465.

—— 1935. Ear exostoses. *Am. J. phys. Anthrop.*, Washington, **20**: 489–490.

—— 1939. Normal micro- and macrocephaly in America. *Am. J. phys. Anthrop.*, Washington, **25**: 1–91.

—— 1940. Mandibular and maxillary hyperostoses. *Am. J. phys. Anthrop.*, Washington, **27**: 1–68.

HUDSON, E. H. 1946. *Treponematosis.* iv, 9–122 +2 pp. New York, Oxford UP.

HUG, E. 1956. *Die Anthropologische Sammlung im Naturhistorischen Museum Bern.* 55 pp. Bern.

HUMPHREYS, H. 1951. Dental evidence in archaeology. *Antiquity*, Gloucester, **25**: 16–18.

HUNT, E. E. & GLEISER, I. 1955. The estimation of age and sex of preadolescent children from bones and teeth. *Am. J. phys. Anthrop.*, Washington, (n.s.) **13**: 479–488.

HUXLEY, T. H. 1862. Notes upon the human remains from the valley of the Trent, and from the Heathery Burn Cave, Durham. *The Geologist*, London, **5**: 201–204.

IMAI, Y. 1970. An electron microscope study on antique human bones and Pleistocene fossil bones. *Proc. VIII int. Congr. anthrop. ethnol. Sci.*, Toyko, **1**: 118.

IMRIE, J. A. & WYBURN, G. M. 1958. Assessment of age, sex and height from immature human bones. *Brit. Med. J.*, London, **1**: 128–131.

IVANOVSKY, A. 1923. Physical modifications of the population of Russia under famine. *Am. J. phys. Anthrop.*, Washington, **6**: 331–353.

JACKSON, W. P. U., DOWDLE, E. & LINDER, G. C. 1958. Vitamin-D-resistant osteomalacia. *Brit. Med. J.*, London, **1**: 1269–1274.

JANSENS, P. A. 1970. *Palaeopathology. Diseases and Injuries of Pre-historic Man.* 170 pp. London, Baker.

JARCHO, S. (ed.) 1966. *Human Palaeopathology.* 182 pp. New Haven, Yale UP.

JIT, I. & SINGH, S. 1956. Estimation of age from clavicles. *Ind. J. Med. Res.*, Calcutta, **44**: 137–155.

JOHNSTON, F. E. 1962. Growth of the long bones of infants and young children at Indian Knoll. *Am. J. phys. Anthrop.*, Washington, **20**: 249–254.

—— & SNOW, C. E. 1961. The reassessment of the age and sex of the Indian Knoll skeletal population: demographic and methodological aspects. *Am. J. phys. Anthrop.*, Washington, **19**: 237–244.

JONES, E. W. A. H. 1931–32. Studies in achondroplasia. *J. Anat. Lond.*, **66**: 565–577.

KADANOFF, D. & MUTAFOV, S. 1968. Über die Variationen der typisch lokalisierten überzähligen Knochen und Knochenfortsätze des Hirnschädels beim Menschen. *Anthrop. Anz.*, **31**: 28–39.

KAUFMANN, H. 1945. Un cas de parietal biparti chez un crâne ancien (Gland, Vaud). *Bull. Soc. Suisse Anthrop. Ethnol.*, **21**: 4–7.

KEEN, J. A. 1950. A study of the differences between male and female skulls. *Am. J. phys. Anthrop.*, Washington, (n.s.) **8**: 65–79.

KEITH, A. 1913a. Abnormal crania – achondroplastic and acrocephalic. *J. Anat. Lond.*, **47**: 189–206.

—— 1913b. Problems relating to the teeth of the earlier forms of prehistoric man. *Proc. roy. Soc. Med.*, London, **6**: 103.

—— 1929. *The Antiquity of Man.* 2 vols. 753 pp. London.

—— 1931. *New Discoveries Relating to the Antiquity of Man.* 512 pp. London.

KEITH, L. B. 1940. Composite odontoma. *J. Am. Dent. Assoc.*, Chicago, **27**: 1479–1480.

KELLEY, M. A. & EL-NAJJAR, M. Y. 1980. Natural variation and differential diagnosis of skeletal changes in tuberculosis. *Am. J. phys. Anthrop.*, Washington, **52**: 153–167.

KENDRICK, G. S. & RISINGER, H. L. 1967. Changes in the anteroposterior dimensions of the human male skull during the third and fourth decade of life. *Am. J. phys. Anthrop.*, Washington, **159**: 77–82.

KERLEY, E. R. 1965. The microscopic determination of age in human bone. *Am. J. phys. Anthrop.*, Washington, **23**: 149–164.

—— & UBELAKER, D. H. 1978. Revisions in the microscopic method of estimating age at death in human cortical bone. *Am. J. phys. Anthrop.*, Washington, **49**: 545–546.

KISZELY, I. 1970. On the peculiar custom of the artificial mutilation of the foramen occipitale magnum. *Acta Arch. Acad. Sci. Hung.*, **22**: 301–321.

—— 1974. On the possibilities and methods of the chemical determination of sex from bones. *Ossa*, **1**: 51–62.

KLATSKY, M. 1956. The incidence of six anomalies of the teeth and jaws. *Hum. Biol.*, Baltimore, **28**: 420–428.

KNIP, A. S. 1971. The frequencies of non-metrical variants in Tellem and Nokara skulls from the Mali Republic. *Proc. K. ned. Akad. Wet.*, C **74**: 422–443.

KOBAYASHI, K. 1964. Estimation of age at death from pubic symphysis for the prehistoric human remains in Japan. *Zinruigaku Zassi*, **72**: 1–13.

KOUT, M., VACIKOVA, A. & STLOUKAL, M. 1965. An attempt to assess blood groups in paleoanthropological material. *Anthropologie, Paris*, **2**: 49–58.

KOWALSKI, C. J. 1972. A commentary on the use of multivariate statistical methods in anthropometric research. *Am. J. phys. Anthrop.*, Washington, **36**: 119–132.

KRAMBERGER, K. G. 1906. *Der Diluviale Mensch von Krapina in Kroatien*. 277 pp., Wiesbaden.

KROGMAN, W. M. 1938a. The skeleton talks. *Sci. Amer.*, New York.

—— 1938b. The role of urbanization in the dentitions of various population groups. *Z. Rassenk.*, Stuttgart, **7**: 41–72.

—— 1939. A guide to the identification of human skeletal material. *F.B.I. Law Enforcement Bull.*, **8**: 8–29.

—— 1946. The skeleton in forensic medicine. *Proc. Inst. Med.*, Chicago, **16**: 154.

KROGSTAD, O. 1972. The maxillary prognathism in Greenland Eskimo – a roentgen craniometric study. *Z. Morph. Anthrop.*, **64**: 279–307.

LAMBERT, J. B., SZPUNAR, C. B. & BUIKSTRA, J.E. 1979. Chemical analysis of excavated human bone from Middle and Late Woodland sites. *Archaeometry*, **21**: 115–129.

LARNACH, S. L. 1974a. Frontal recession and artificial deformation. *Archaeol. phys. Anthrop. Oceania*, **9**: 214–216.

—— 1974b. An examination of the use of discontinuous cranial traits. *Archaeol. phys. Anthrop. Oceania*, **9**: 217–225.

LASKER, G. W. 1947. Penetrance estimated by the frequency of unilateral occurrences and by discordance in monozygotic twins. *Hum. Biol.*, Baltimore, **19**: 217–230.

—— 1950. Genetic analysis of racial traits of the teeth. *Cold Spring Harbor Symposia on Quantitative Biology*, **15**: 191–203.

LASKER, H. L. 1946. Migration and physical differentiation. *Am. J. phys. Anthrop.*, Washington, **4**: 273–300.

LAUGHLIN, W. S. 1948. Appendix F. Preliminary tests for the presence of blood-group substances in Tepexpan man. *Tepexpan man*: 132–135. New York (Viking Fund Publs Anthrop. **11**).

—— & JØRGENSEN, J. B. 1956. Isolate variation in Greenlandic Eskimo. *Acta Genet.*, **6**: 3–12.

LAURENCE, K. M. 1958. The natural history of hydrocephalus. *Lancet*, London, **1958**: 1152–1154.

LAVELLE, C. L. B. & MOORE, W. J. 1969. Alveolar bone resorption in Anglo-Saxon and seventeenth century mandibles. *J. periodont. Res.*, **4**: 70–73.

LAW, W. A. 1950. Surgical procedures in the treatment of chronic arthritis of the spine. *Ann. Roy. Coll. Surg. Eng.*, **6**: 56–69.

LE DOUBLE, A. F. 1903. *Traité des variations des os du crâne de l'homme*. 400 pp. Paris.

LEECHMAN, D. 1931. Technical methods in the preservation of anthropological museum specimens. *Bull. Nat. Mus. Canada*, Ottawa, **67**: 127–158.

LE GROS CLARK, W. E. 1955. *The Fossil Evidence for Human Evolution*. 181 pp. Chicago.

—— 1957. Re-orientations in physical anthropology. *In* Roberts, D. F. & Weiner, J. S. (eds), *The scope of physical anthropology and its place in academic studies*: 1–6. (Publd for Soc. Study hum. Biol. by Wenner-Gren Foundation).

LEIGH, R. W. 1925. Dental pathology of Indian tribes of varied environmental and food conditions. *Am. J. phys. Anthrop.*, Washington, **8**: 179–199.

LENGYEL, I. 1971. Chemico-analytical aspects of human bone finds from the 6th century "Pannonian" cemeteries. *Acta Arch. Acad. Sci. Hung.*, **23**: 139–166.

LINNÉ, S. 1943. Humpbacks in ancient America. *Ethnos*, Mexico, **8**: 161–186.

LISOWSKI, F. P. 1955/56. The cremations from the Culdoich, Leys and Kinchyle sites. *Proc. Soc. Antiq. Scot.*, **89**: 83–90.

—— 1956. The cremations from Barclodiad y Gawres. *In* Powell, T. G. E. & Daniel, G. E., *Barclodiad y Gawres.* Liverpool.

—— 1967. Prehistoric and early historic trepanation. *In* Brothwell, D. R. & Sandison, A. T. (eds), *Diseases in Antiquity:* 651–672. Springfield.

LOVEJOY, C. O., BURSTEIN, A. H. & HEIPLE, K. G. 1976. The biomechanical analysis of bone strength: a method and its application to platycnemia. *Am. J. phys. Anthrop.*, Washingon, **44**: 489–506.

LUMHOLTZ, C. & HRDLIČKA, A. 1898. Marked human bones from a prehistoric Tarasco Indian burial place in the state of Michoacon, Mexico. *Bull. Am. Mus. Nat. Hist.*, New York, **10**: 61–79.

LUNT, D. A. 1954. A case of taurodontism in a modern European molar. *Dent. Rec.*, London, **1954**: 307–312.

—— 1969. An Odontometric Study of Mediaeval Danes. *Acta Odont. Scand.* **27** (Suppl. 55). 173 pp.

MACADAM, R. F. & SANDISON, A. T. 1969. The electron microscope in palaeopathology. *Med. Hist.*, **13**: 81–85.

MCCOWN, T. D. & KEITH, A. 1939. *The Stone Age of Mount Carmel*, **2**. 390 pp. Oxford.

MACCURDY, G. G. 1923. Human skeletal remains from the highlands of Peru. *Am. J. phys. Anthrop.*, Washington, **6**: 217–329.

MACDONELL, W. R. 1906. A second study of the English skull with special reference to Moorfields crania. *Biometrika*, Cambridge, **5**: 86–104.

MCKERN, T. W. & STEWART, T. D. 1957. *Skeletal Age Changes in Young American Males.* 179 pp. Natick, Mass. (Technical report. Headquarters Quartermaster Research and Development Command.)

MAHALANOBIS, P. C. 1936. On the generalized distance in statistics. *Proc. Nat. Inst. Sci. India*, **12**: 49–55.

MANTLE, J. A. 1959. Non-tuberculous infections of the spine. *In* Nassim, R. & Burrows, H. J. (eds), *Modern Trends in Diseases of the Vertebral Column*: 142–154. London, Butterworth.

MARIJA, GIMBUTAS 1956. *The Prehistory of Eastern Europe, Part I: Mesolithic, Neolithic and Copper Age cultures in Russia and the Baltic areas.* 241 pp. Cambridge, Mass. Peabody Museum Publications.

MARSHALL, D. S. 1955. Precondylar tubercle incidence rates. *Am. J. phys. Anthrop.*, Washington, (n.s.) **13**: 147–151.

MARSHALL, W. A. 1968. Problems in relating the presence of transverse lines in the radius to the occurrence of disease. *In* Brothwell, D. R. (ed.), *The skeletal biology of earlier human populations*: 245–261. London, Pergamon.

MARTIN, C. P. 1935. *Prehistoric Man in Ireland.* 184 pp. London.

MASALI, M. 1964. Dati sulla variabilita morfometrica e ponderale degli ossicini dell'udito nell'uomo. *Arch. Ital. Anat. Embryol.*, Florence, **69**: 435–446.

—— 1972. Body size and proportions as revealed by bone measurements and their meaning in environmental adaptation. *J. hum. Evol.*, **1**: 187–197.

MASSET, C. 1973. La démographie des populations inhumées. Essai de paléodemographie. *Homme*, **13**: 95–131.

MATSON, G. A. 1936. A procedure for the serological determination of blood relationship in ancient and modern peoples with special reference to the American Indians. II. Blood grouping in mummies. *J. Immunol.*, Baltimore, **30**: 459–470.

MEHNERT, E. 1892/93. Catalog der anthropologischen Sammlung des Anatomischen Instituts, der Universitäts Strasburg. I. E. *Archiv. Anthrop.*, Braunschweig, **1892/93**: 21.

MELLANBY, E. 1934. *Nutrition and Disease.* 171 pp. Edinburgh.

MELLANBY, M. 1934. *Diet and the teeth; an experimental study. Part III: The effect of diet on dental structure and disease in man.* Special Report Series – Medical Research Council. 191 pp. London, H.M.S.O.

MERBS, C. F. 1967. Cremated human remains from Point of Pines, Arizona. A new approach. *Am. Antiq.*, **32**: 498–506.

MERCHANT, V. L. & UBELAKER, D. H. 1977. Skeletal growth of the protohistoric Arikara. *Am. J. phys. Anthrop.*, Washington, **46**: 61–72.

RAO, C. R. 1948. The utilization of multiple measurements in problems of biological classification. *J. Roy. Statist. Soc.*, London, **10**: 159.

READER, R. 1974. New evidence for the antiquity of leprosy in early Britain. *J. arch. Sci.*, **1**: 205–207.

REDFIELD, A. 1970. A new aid to ageing immature skeletons: development of the occipital bone. *Am. J. phys. Anthrop.*, Washington, **33**: 207–220.

REQUENA, A. 1946. Evidencia de tuberculosis en la America precolombina. *Acta Venozolana*, **1**: 1–20.

REYNOLDS, E. L. 1945. The bony pelvic girdle in early infancy. *Am. J. phys. Anthrop.*, Washington, (n.s.) **3**: 231–254.

—— 1947. The bony pelvis in prepuberal childhood. *Am. J. phys. Anthrop.*, Washington, (n.s.) **5**: 165–200.

RIESENFELD, A. 1956. Multiple infra-orbital, ethmoidal and mental foramina in the races of man. *Am. J. phys. Anthrop.*, Washington, (n.s.) **14**: 85–100.

RIGHTMIRE, G. P. 1972. Cranial measurements and discrete traits compared in distance studies of African negro skulls. *Hum. Biol.*, **44**: 263–276.

RILEY, M. J., ANSELL, B. M. & BYWATERS, E. G. L. 1971. Radiological manifestations of ankylosing spondylitis according to age at onset. *Ann. rheum. Dis.*, **30**: 138–148.

RISDON, D. L. 1939. A study of the cranial and other human remains from Palestine excavated at Tell Duweir (Lachish) by the Wellcome-Marston Archaeological Research Expedition. *Biometrika*, Cambridge, **31**: 99–166.

RITCHIE, P. R. & PUGH, J. 1963. Ultra-violet radiation and excavation. *Antiquity*, Cambridge, **37**: 259–263.

RITCHIE, W. A. 1952. Paleopathological evidence suggesting pre-Columbian tuberculosis in New York State. *Am. J. phys. Anthrop.*, Washington, (n.s.) **10**: 305–317.

—— & WARREN, S. L. 1932. The occurrence of multiple bony lesions suggesting myeloma in the skeleton of a pre-Columbian Indian. *Amer. J. Roentgenol.*, New York, **28**: 622–628.

RIXON, A. 1976. *Fossil Animal Remains.* 304 pp. London.

ROBINSON, J. T. 1952. Some hominid features of the ape-man dentition. *J. dent. Assoc. S. Afr.*, **7**: 102–113.

ROCHE, A. F. 1953. Increase in cranial thickness during growth. *Hum. Biol.*, Baltimore, **25**: 81–92.

RUFFER, A. 1920. Study of abnormalities and pathology of ancient Egyptian teeth. *Am. J. phys. Anthrop.*, Washington, **3**: 335–382.

RUSCONI, C. 1940. Parietal múltiple en un cráneo indígena de Mendoza. *Semana Médica*, Buenos Aires, **47** (24). 7 pp.

—— 1946. La piorrea en los indígenas prehispánicos de Mendoza. *Rev. odontal.*, Paris, **34**: 118–121.

RYDER, M. L. 1969. *Animal Bones in Archaeology.* xxiv + 65 pp., 5 tabs. Oxford.

SAGER, P. 1969. *Spondylosis cervicalis. Årsberetn. Kφbenhavns Univs Medic.-Hist. Inst. Mus.*, **1968-69**: [185]–224. (Thesis summary in English and Danish).

SALAMA, N. & HILMY, A. 1951. An ancient Egyptian skull and a mandible showing cysts. *Brit. dent. J.*, London, **90**: 17–18.

SATINOFF, M. I. 1972. The medical biology of the early Egyptian populations from Asswan, Assyut and Gebelen. *J. hum. Evol.*, **1**: 247–257.

SAUNDERS, S. R. & POPOVICH, F. 1978. A family study of two skeletal variants: atlas bridging and clinoid bridging. *Am. J. phys. Anthrop.*, Washington, **49**: 193–204.

SAVARA, B. S. 1965. A method of measuring facial bone growth in three dimensions. *Hum. Biol.*, **37**: 245–255.

SCHAEFER, U. 1955. Demographische beobachtungen an der wikingerzeitlichen Bevölkerung von Haithabu, und Mitteilung einiger pathologischer befunde an den Skeletten. *Z. Morph. Anthr.*, Stuttgart, **47**: 221–228.

SCHLAGINHAUFEN, O. 1925. *Die menschlichen Skelettreste aus der Steinzeit des Wauwilersees (Luzern) und ihre Stellung zu andern anthropologischen Funden aus der Steinzeit.* Zürich. (Quoted by Sigerist, 1951).

SCHMID, E. 1972. *Atlas of animal bones for prehistorians, archaeologists and quarternary geologists.* 159 pp. Amsterdam, Elsevier.

SCHOENINGER, M. J. 1979. Dietary reconstruction at Chalcatzingo, a Formative Period site in Morelos, Mexico, *Tech. Rep. Mus. Anthrop. Univ. Mich.*, **9**. 97 pp.

SCHOFIELD, G. 1959. Metric and morphological features of the femur of the New Zealand Maori. *J. Roy. Anthrop. Inst.*, London, **89**: 89–105.

SCHOUR, I. & MASSLER, M. 1941. The development of the human dentition. *J. Am. dent. Assoc.*, Chicago, **28**: 1153–1160.

SCHULTZ, A. H. 1930. The skeleton of the trunk and limbs of higher primates. *Hum. Biol.*, Baltimore, **2**: 303–456.

—— – 1956. The occurrence and frequency of pathological and teratological conditions and of twinning among non-human primates. *Primatologia*, Basel, **1**: 965–1014.

SCHWALBE, G. 1903. Über geteilte Scheitelbeine. *Z. Morph. Anthr.*, Stuttgart, **6**: 361–434.

SELIGMAN, C. G. 1912. A cretinous skull of the Eighteenth Dynasty. *Man*, London, **12**: 17–18.

SELTZER, C. C. 1937. A critique of the coefficient of racial likeness. *Am. J. phys. Anthrop.*, Washington, **23**: 101–109.

SERGI, S. 1937. Ossicini fontanellari della regione del lambda nel cranio di Saccopastore e nei crani neandertaliani. *Riv. Antrop.*, Roma, **30**: 3–14.

——, ASCENZI, A. & BONUCCI, E. 1972. Torus palatinus in the Neanderthal Circeo I skull. A histologic, microradiographic and electron microscopic investigation. *Am. J. phys. Anthrop.*, Washington, **36**: 189–198.

SHAPIRO, H. L. 1939. *Migration and Environment*. 594 pp. London.

SHELDON, W. H. 1954. *Atlas of Men*. 357 pp. New York.

SIGERIST, H. E. 1951. *A History of Medicine. I. Primitive and Archaic Medicine*. 564 pp. New York.

SIMMONS, D. C. 1957. The depiction of gangosa on Efik-Ibibio masks. *Man*, London, **57**: 17–20.

SINGER, R. 1953a. Artificial deformation of teeth. *S. Afr. J. Sci.*, Cape Town, **50**: 116–122.

—— 1953b. Estimation of age from cranial suture closure. *J, For. Med.*, Cape Town, **1**: 52–59.

—— 1958. The Boskop 'race' problem. *Man*, London, **58**: 193–232.

SISSONS, S. & GROSSMAN, J. D. 1953. *The Anatomy of the Domestic Animals*. 972 pp. London.

SJØVOLD, T. 1974. Some aspects of physical anthropology on Gotland during Middle Neolithic times. *In* Janzon, G. O., *Gotlands Mellanneolitiska Gravar*: 176–211. Stockholm, Almquist & Wiksell.

—— 1977. Non-metrical divergence between skeletal populations. *Ossa*, **4** (Suppl. 1). 133 pp.

——, SWEDBORG, I. & DIENER, L. 1974. A pregnant woman from the Middle Ages with exostosis multiplex. *Ossa*, **1**: 3–23.

SMITH, H. O. 1912. A study of pygmy crania, based on skulls found in Egypt. *Biometrika*, Cambridge, **8**: 262–266.

SMITH, M. 1959a. Blood grouping of the remains of Swedenborg. *Nature, Lond.*, **184**: 867–869.

—— 1959b. Discussion. *In: Medical Biology and Etruscan Origins:* 150–152. London.

—— 1960. The blood groups of the ancient dead. *Science*, Washington, **131**: 699–702.

SMITH, SIR S. & FIDDES, F. S. 1955. *Forensic Medicine*, 10th ed. 644 pp. London.

SMYTH, K. C. 1933. Some notes on the dentitions of the Anglo-Saxon skulls from Bidford-on-Avon, with special reference to malocclusion. *Trans. brit. Soc. Orthodont.*, London, **1933**: 1–21.

SOGNNAES, R. F. 1956. Histological evidence of developmental lesions in teeth originating from palaeolithic, pre-historic and ancient times. *Am. J. Path.*, Boston, **32**: 547–577.

SPENCE, T. F. 1967. The anatomical study of cremated bone fragments from archaeological sites. *Proc. Prehist. Soc.*, Cambridge, **33**: 70–83.

—— & TONKINSON, T. 1969. The application of ultrasonic cleaning in osteological studies. *J. Sci. Technol.*, **15**: 47–50.

SPITTEL, R. L. 1923. *Framboesia Tropica*. 59 pp. London.

STALLWORTHY, J. A. 1932. A case of enlarged parietal foramina associated with metopism and irregular synostosis of the coronal suture. *J. Anat. Lond.*, **67**: 168–174.

STEELE, D. G. & MCKERN, T. W. 1969. A method for assessment of maximum long bone length and living stature from fragmentary long bones. *Am. J. phys. Anthrop.*, Washington, **31**: 215–227.

STEIDA, L. 1894. Über die verschiedenen Formen der Sog. queren Gaumennaht (sutura palatina transversa). *Archiv. Anthrop.*, Braunschweig, **22**: 1–12.

STEINBOCK, R. T. 1976. *Paleopathological Diagnosis and Interpretation.* 423 pp. Thomas, Springfield.

STERN, C. 1950. *Principles of Human Genetics.* 617 pp. San Francisco.

STEVENSON, P. H. 1929. On racial differences in stature long-bone regression formulae, with special references to stature reconstruction formulae for the Chinese. *Biometrika*, Cambridge, **21**: 303–321.

—— 1930. A convenient anthropological record form for field workers. *Man*, London, **30**: 78–81.

STEWART, T. D. (ed.) 1947. *Hrdlička's Practical Anthropometry*, 3rd ed. 230 pp. Philadelphia, Wistar Institute.

—— 1950. Pathological changes in South American Indian skeletal remains. *Smithson. Inst. Am. Ethn. Bull.*, Washington, **6**: 49–52.

—— 1954. Sex determination in the skeleton by guess and by measurement. *Am. J. phys. Anthrop.*, Washington, (n.s.) **12**: 385–392.

—— 1956a. Examination of the possibility that certain skeletal characters predispose to defects in the lumbar neural arches. *Clinical Orthopaedics*, Philadelphia, **8**: 44–60.

—— 1956b. Skeletal remains from Xochicalco, Morelos. *Estudios Antropológicos México, D.F.* **1956**: 131–156.

—— 1956c. Significance of osteitis in ancient Peruvian trephining. *Bull. Hist. Med.*, **30**: 293–320.

—— 1957. Distortion of the pubic symphyseal surface in females and its effect on age determination. *Am. J. phys. Anthrop.*, Washington, (n.s.) **15**: 9–18.

—— 1958a. Stone Age skull surgery, a general review with emphasis on the New World. *Smithson. Inst. Rept.* (**1957**): 469–491.

—— 1958b. The rate of development of vertebral osteo-arthritis in American Whites and its significance in skeletal age identification. *The Leech*, Johannesburgh, **28**: 144–151.

—— 1966. Some problems in human palaeopathology. *In* Jarcho, S. (ed.), *Human Palaeopathology*: 43–55. New Haven, Yale U.P.

—— 1975. Cranial dysraphism mistaken for trephination. *Am. J. phys. Anthrop.*, Washington, **42**: 435–438.

—— & SPOEHR, A. 1952. Evidence on the palaeopathology of yaws. *Bull. Hist. Med.* **26**: 538–553.

STOUT, S. D. 1978. Histological structure and its preservation in ancient bone. *Curr. Anthrop.*, **19**: 601–603.

STRAUS, W. L. & CAVE, A. J. E. 1957. Pathology and posture of Neanderthal man. *Q. Rev. Biol.*, Baltimore, **32**: 348–363.

SULLIVAN, L. R. 1922. The frequency and distribution of some anatomical variations in American crania. *Anthrop. Pap. Am. Mus.*, New York, **23**: 207–258.

SUZUKI, H. 1975. A case of microcephaly in an Aeneolithic Yayoi Period population in Japan. *Bull. natn. Sci. Mus. Toyko* D, **1**: 1–10.

—— et al. 1956. *Mediaeval Japanese Skeletons from the Burial Site at Zimokuza, Kamakura City:* 172–194. Tokyo.

SWEDLUND, A. C. & ARMELAGOS, G. J. 1976. *Demographic Anthropology.* 70 pp. Dubuque, Brown.

SZPUNAR, C. B., LAMBERT, J. B. & BUIKSTRA, J. E. 1978. Analysis of excavated bone by atomic absorption. *Am. J. phys. Anthrop.*, Washington, **48**: 199–202.

TAPPEN, N. 1955. A comparative functional analysis of primate skulls by the split-line technique. *In* Gavan, J. A. (arr.), *The nonhuman primates and human evolution*: 42–59. Detroit, Wayne University.

TAYLOR, R. M. S. 1972. Terminal hypoplasia in tooth roots. *Bull. Grpmt int. Rech. scient. Stomat.*, **15**: 97–106.

TELKKÄ, A. 1950. On the prediction of human stature from long-bones. *Acta Anatomica*, Basle, **9**: 103–117.

THIEME, F. P. 1957. Sex in negro skeletons. *J. For. Med.*, Cape Town. **4**: 72–81.

——, OTTEN, C. M. & SUTTON, H. E. 1956. A blood typing of human skull fragments from the Pleistocene. *Am. J. phys. Anthrop.*, Washington, (n.s.) **14**: 437–443.

——, —— & WHEELER, A. H. 1957. *Biochemical and immunological identification of human remains. (Report of research.)* 18 pp. Natick, Mass.

—— & SCHULL, W. J. 1957. Sex determination from the skeleton. *Hum. Biol.*, Baltimore, **29**: 242–273.

THOMA, K. H. 1946. *Oral Pathology*. 1328 pp. London.

THOMPSON, A. 1899. The sexual differences of the foetal pelvis. *J. Anat. Lond.*, **33**: 359–380.

TILDESLEY, M. L. 1931. Bones and the excavator. *Man*, London, **31**: 100–103.

—— 1956. A critical survey of techniques for the measurements of cranial capacity. *J. Roy. Anthrop. Inst.*, London, **83**: 182–193.

TOBIAS, P. V. 1960. The Kanam Jaw. *Nature, Lond.*, **195**: 946–947.

TODD, T. W. 1920. Age changes in the pubic bone. I. The Male White pubis. *Am. J. phys. Anthrop.*, Washington, **3**: 285–334.

—— 1921. Age changes in the pubic bone. II. Pubis of Male Negro-White hybrid; III. Pubis of White Female; IV. Pubis of Female Negro-White hybrid. *Am. J. phys. Anthrop.*, Washington, **4**: 1–17.

—— & LYON, D. W. 1924. Endocranial suture closure. Part I: Adult males of White stock. *Am. J. phys. Anthrop.*, Washington, **7**: 325–384.

—— —— 1925. Cranial suture closure. Part II: Ectocranial closure in adult males of White stock. *Am. J. phys. Anthrop.*, Washington, **8**: 23–71.

TORGERSEN, J. 1951. The developmental genetics and evolutionary meaning of the metopic suture. *Am. J. phys. Anthrop.*, Washington, (n.s.) **9**: 193–210.

—— 1954. The occiput, the posterior cranial fossa and the cerebellum. *In* Jansen, J. & Brodal, A., *Aspects of cerebellar anatomy*: 396–418. Oslo.

TOWNSLEY, W. 1946. Platymeria. *J. Path. Bact.*, London, **58**: 85–88.

TRATMAN, E. K. 1950. A comparison of the teeth of people. *Dent. Rec.*, London, **70**: 31–53; 63–88.

TREVOR, J. C. 1950. Notes on the human remains of Romano-British date from Norton, Yorks. *In* Whitley, H., *The Roman Pottery at Norton, East Yorkshire*. (Roman Malton and District Report no. 7). Leeds.

—— 1953. Race crossing in man. The analysis of metrical characters. *Eugen. Lab. Mem.*, London, **36**. 45 pp.

—— 1967. Anthropometry. *In: Chambers' Encyclopaedia*, new revised edn, **1**: 469–474. London.

TRINKAUS, E. 1975. Squatting and Neandertals: a problem in the behavioural interpretation of skeletal morphology. *J. archaeol. Sci.*, **2**: 327–351.

—— 1978. Bilateral asymmetry of human skeletal non-metric traits. *Am. J. phys. Anthrop.*, Washington, **49**: 315–318.

TROTTER, M. & GLESER, G. C. 1952. Estimation of stature from long-bones of American Whites and Negroes. *Am. J. phys. Anthrop.*, Washington, (n.s.) **10**: 463–514.

—— —— 1958. A re-evaluation of estimation of stature based on measurements of stature taken during life and long-bones after death. *Am. J. phys. Anthrop.*, Washington, (n.s.) **16**: 79–123.

TULSI, R. S. 1972. Vertebral column of the Australian aborigine: selected morphological and metrical features. *Z. Morph. Anthrop.*, **64**: 117–144.

TWIESSELMANN, F. & BRABANT, H. 1967. Les dents et les maxillaires de la population d'age Franc de Coxyde (Belgique). *Bull. Grpmt int. Rech. scient. Stomat.*, **10**: 5–180.

UBELAKER, D. 1974. Reconstruction of demographic profiles from ossuary skeletal samples. A case study from the Tidewater Potomac. *Smithson. Contrib. Anthrop.*, **18**. 79 pp.

——, PHENICE, T. W. & BASS, W. M. 1969. Artificial interproximal grooving of the teeth in American Indians. *Am. J. phys. Anthrop.*, Washington, **30**: 145–150.

UCKO, P. J. 1969. Ethnography and archaeological interpretation of funerary remains. *Wld Archaeol.*, London, **1**: 262–280.

ULLRICH, H. & WEICKMANN, F. 1965. Prähistorische trepanationen und ihre Abgrenzung gegen andere Schädeldachdefekte. *Anthrop. Anz.*, **29**: 261–272.

VALLOIS, H. V. 1936. La carie dentaire à la chronologie des hommes pré-historiques. *Anthropologie*, Paris, **46**: 201–212.

—— 1938. Les méthodes de mensuration de la platycnemie; Étude critique. *Bull. Soc. Anthrop.*, Paris, **9**: 97–108.

—— 1946. L'omoplate humaine. *Bull. Soc. Anthrop.*, Paris, **7**: 16–99.

—— 1957. Le poids comme caractère sexuel des os longs. *Anthropologie*, Paris, **61**: 45–69.

VAN GERVEN, D. P. 1973. Thickness and area measurements as parameters of skeletal involution of the humerus, femur and tibia. *J. Geront.*, **28**: 40–45.

VAN VARK, G. N. 1970. *Some statistical procedures for the investigation of prehistoric human skeletal material*. 124 pp. Groningen, Rijksuniversiteit.

—— 1974. The investigation of human cremated skeletal material by multivariate statistical methods. I. Methodology. *Ossa*, **1**: 63–95.

VARE, A. M. 1974. Angle of femoral torsion. *Hum. Biol.*, **46**: 233–238.

VARGAS, L. 1974. Caracteres craneanos discontinuous en la población de Tlatilco, Mexico. *An. Anthrop.*, **11**: 307–328.

VERGER-PRATOUCY, J.-C. 1968. *Recherches sur les Mutilations maxillo-dentaires préhistoriques*. M.D. thesis, Univ. of Bordeaux.

VIDIĆ, B. 1968. The structure of the palatum osseum and its toral overgrowths. *Acta anat.*, **71**: 94–99.

VIRTAMA, P. & HELELÄ, T. 1969. Radiographic measurements of cortical bone. *Acta radiol.* Suppl. **293**.

VLČEK, E., KOMINEK, J., ANDRIK, P. & BÍLY, B. 1975. Proposal of unification in documenting and determining the dental age on skeletal material. *Scripta medica*, **48**: 299–311.

WAHBY, W. 1903/5. Abnormal nasal bones. *J. Anat. Lond.*, **38**: 49–51.

WALDRON, H. A., 1973. Lead poisoning in the ancient world. *Med. Hist.*, **17**: 392–399.

——, MACKIE, A. & TOWNSHEND, A. 1976. The lead content of some Romano-British bones. *Archaeometry*, **8**: 221–227.

WANKEL, H. 1883. Die Funde in der Býčiskála-Höhle. *Archiv. Anthrop.*, Braunschweig, **14**: 45–48.

WARD, R. H. & WEISS, K. M. 1976. *The Demographic Evolution of Human Populations*. 158 pp. London, Academic Press.

WASHBURN, S. L. 1948. Sex differences in the pubic bone. *Am. J. phys. Anthrop.*, Washington, (n.s.) **6**: 199–208.

WATERSTON, D. 1927. A stone cist and its contents found at Piekie Farm, near Boarhills, Fife. *Proc. Soc. Antiq. Scot.*, **61**: 30–44.

WATSON, E. H. & LOWREY, G. H. 1951. *Growth and Development of Children*. 260 pp. Chicago.

WEAVER, D. S. 1980. Sex differences in the ilia of a known sex and age sample of fetal and infant skeletons. *Am. J. phys. Anthrop.*, Washington, **52**: 191–195.

WEINER, J. S. 1951. Cremated remains from Dorchester. *In: Excavations at Dorchester, Oxon:* 129–141. Oxford, Ashmolean Museum.

——, OAKLEY, K. P. & LE GROS CLARK, W. E. 1953. The solution of the Piltdown problem. *Bull. Br. Mus. nat. Hist.*, London, (Geol.) **2**: 141–146.

—— —— —— *et al.* 1955. Further contributions to the solution of the Piltdown problem. *Bull. Br. Mus. nat. Hist.*, London, (Geol) **2**: 227–287.

—— & THAMBIPILLAI, V. 1952. Skeletal maturation of West African Negroes. *Am. J. phys. Anthrop.*, Washington, (n.s.) **10**: 407–418.

WEINMANN, J. P. & SICHER, H. 1947. *Bone and Bones. Fundamentals of Bone Biology*. 464 pp. St Louis.

WEISS, D. L. & MØLLER-CHRISTENSEN, V. 1971. An unusual case of tuberculosis in a medieval leper. *Årsberetn. Københavns Univs Medic.-Hist. Inst. Mus.*, **1971**. 10 pp.

WEISS, K. M. 1972. On the systematic bias in skeletal sexing. *Am. J. phys. Anthrop.*, Washington, **37**: 239–250.

WELCKER, H. 1887. Cribra orbitalis. *Archiv. Anthrop.*, Braunschweig, **17**: 1–18.

WELLS, C. 1960. A study of cremation. *Antiquity*, Gloucester, **34**: 29–37.

—— 1962. A possible case of leprosy from a Saxon cemetery at Beckford. *Med. Hist.*, London, **6**: 383.

—— 1963. The radiological examination of human remains. *In: Science in Archaeology*: 401–412. London.

—— 1964. *Bones, Bodies and Disease*. 288 pp. London, Thames & Hudson.

—— 1973. A palaeopathological rarity in a skeleton of Roman date. *Med. Hist.*, **17**: 399–400.

—— 1974. Osteochondritis dissecans in ancient British skeletal material. *Med. Hist.*, **18**: 365–369.

WERNER, A. E. A. 1958. Technical notes on a new material in conservation. *Chron. d'Egypte*, Brussels, **33** (60): 275–278.

WHEELER, R. E. M. 1943. Maiden Castle, Dorset. *Rep. Res. Cttee roy. Soc. Antiq. Lond.*, **12**. 399 pp. Oxford.

WOLPOFF, M. H. 1971. *Metric Trends in Hominid Dental Evolution*. 244 pp. London.

WOO, J.-K. 1948. 'Anterior' and 'Posterior' mediopalatine bones. *Am. J. phys. Anthrop.*, Washington, (n.s.) **6**: 209–223.

—— 1949. Ossification and growth of the human maxilla, premaxilla and palate bone. *Anat. Rec.*, Philadelphia, **105**: 737–762.

WOO, T. L. 1931. On the asymmetry of the human skull. *Biometrika*, Cambridge, **22**: 324–352.

WOODBURY, G. 1936. The use of polymerized vinyl acetate and related compounds in the preservation and hardening of bone. *Am. J. phys. Anthrop.*, Washington, **21**: 449–450.

WOOD JONES, F. 1931. The non-metrical morphological characters of the skull as criteria for racial diagnosis. *J. Anat., Lond.*, **65**: 179–195; 368–378; 438–445.

WOOLLEY, L. 1949. *Digging up the Past*. 122 pp. Harmondsworth. Penguin Books.

WORLD HEALTH ORGANIZATION. 1953. *Expert committee on leprosy*. Technical report series No. **71**. Geneva.

WRIGHT, W. 1903. Pathological conditions in prehistoric skulls. *Birmingham Med. Rev.* (n.s.) **1**: 100–103.

WUNDERLEY, J. 1939. The cranial and other skeletal remains of Tasmanians in collections in the Commonwealth of Australia. *Biometrika*, Cambridge, **30**: 305–337.

ZUHRT, R. 1955. Stomatologische Untersuchungen an Spätmittelalterlichen Funden von Reckkahn. (12–14 Jh.) I. Die Zahnkaries und ihre Folgen. *Dtsche Zahn-, Munds-, und Kieferheilkunds*, **25**: 1–15.

Index

(Numbers in italic indicate figures)